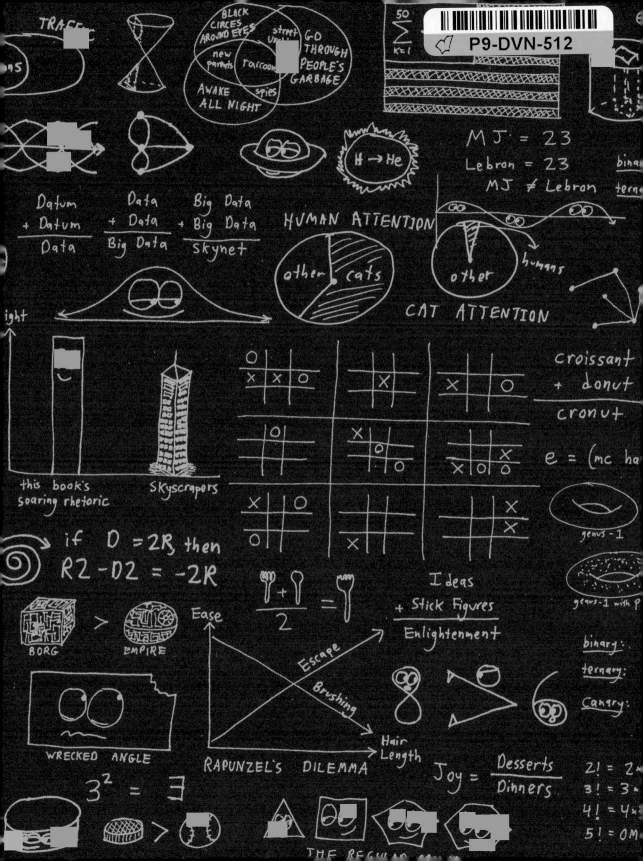

MATH *with*
BAD DRAWINGS

$(1-\lambda)^2 \pi \sqrt{a^2+b^2}$ $|\underline{u}\wedge\underline{v}|=|\underline{u}||\underline{v}|\sin\theta$ $|AB|:|BC|=\lambda:(1-\lambda)^2\pi\sqrt{a^2+b^2}$ $|\underline{u}\wedge\underline{v}|=|\underline{u}||\underline{v}|\sin\theta$

$=\underline{b}-\underline{a}=\lambda(\underline{c}-\underline{a})$ $(\underline{u}\cdot\underline{v})^2+|\underline{u}\wedge\underline{v}|^2=|\underline{u}|^2|\underline{v}|^2$ $\vec{AC}=(\underline{c}-\underline{a}), \vec{AB}=\underline{b}-\underline{a}=\lambda(\underline{c}-\underline{a})$ $(\underline{u}\cdot\underline{v})^2+|\underline{u}\wedge\underline{v}|^2=$

$\underline{c} \quad \lambda=\frac{1}{2}$ $|\underline{u}\wedge\underline{v}|^2=|\underline{u}|^2|\underline{v}|^2\sin^2\theta$ $b=11-\lambda|a+\lambda\underline{c} \quad \lambda=\frac{1}{2}$ $|\underline{u}\wedge\underline{v}|^2=|\underline{u}|^2|\underline{v}|^2\sin$

$\underline{u}\wedge\underline{v}=|\underline{u}||\underline{v}|\sin\theta\,\underline{n} \quad 2\pi\sqrt{a^2+b^2} \quad \underline{b}=\frac{1}{2}(\underline{a}+\underline{c})$ $\underline{u}\wedge\underline{v}=|\underline{u}||\underline{v}|\sin\theta\,\underline{n}$

$\underline{u}\wedge\underline{v}=A\underline{n}$
$A=|\underline{u}||\underline{v}|\sin\theta$
$t+\delta t$
$r(t+\delta t)$

$\underline{d}=\frac{1}{2}(\underline{b}+\underline{c}), \quad \underline{g}=\frac{1}{3}(\underline{a}+\underline{b}+\underline{c})$

area A
$|\underline{u}||\underline{v}|\sin\theta$

$\delta r = r(t+\delta t)\cdot r(t)$
$dt\to 0$
$\frac{dr}{dt}=\lim_{\delta t\to 0}\frac{\delta r}{\delta t}$

$(\underline{u}\cdot\underline{v})^2+|\underline{u}\wedge\underline{v}|^2=|\underline{u}|^2|\underline{v}|^2$

$(u_1u_1+u_2u_2+u_3u_3)^2+(u_2u_3-u_3u_2)^2-$
$(u_3v_1-u_1v_3)^2+(u_1v_2-u_2v_1)^2=$
$=(u_1^2+u_2^2+u_3^2)(v_1^2+v_2^2+v_3^2)$

$(\theta,\varphi)\in[0,2\pi]\times$
$[-\pi/2,\pi/2]$

$x^2+y^2=z^2$

MATH *with* BAD DRAWINGS

ILLUMINATING THE IDEAS THAT SHAPE OUR REALITY

BEN ORLIN

BLACK DOG
& LEVENTHAL
PUBLISHERS
NEW YORK

Black Dog & Leventhal Publishers
Hachette Book Group
1290 Avenue of the Americas
New York, NY 10104
www.hachettebookgroup.com
www.blackdogandleventhal.com

First Edition: August 2018

Black Dog & Leventhal Publishers is an imprint of Running Press, a division of
Hachette Book Group. The Black Dog & Leventhal Publishers name and logo are
trademarks of Hachette Book Group, Inc.

The publisher is not responsible for websites (or their content) that are not owned
by the publisher.

The Hachette Speakers Bureau provides a wide range of authors for speaking events.
To find out more, go to www.HachetteSpeakersBureau.com or call (866) 376-6591.

Print book interior design by Headcase Design.
Library of Congress Control Number: 2018930088
ISBNs: 978-0-316-50903-9 (hardcover); 978-0-316-50902-2 (ebook)
Printed in China
1010
10 9 8 7 6 5 4 3 2 1

For Taryn

CONTENTS

III

PROBABILITY: THE MATHEMATICS OF MAYBE

IV

STATISTICS: THE FINE ART OF HONEST LYING

V

ON THE CUSP: THE POWER OF A STEP

INTRODUCTION

This is a book about math. That was the plan, anyway.

Somewhere, it took an unexpected left turn. Before long, I found myself without cell phone reception, navigating a series of underground tunnels. When I emerged into the light, the book was still about math, but it was about lots of other things, too: Why people buy lottery tickets. How a children's book author swung a Swedish election. What defines a "Gothic" novel. Whether building a giant spherical space station was really the wisest move for Darth Vader and the Empire.

That's math for you. It connects far-flung corners of life, like a secret system of Mario tubes.

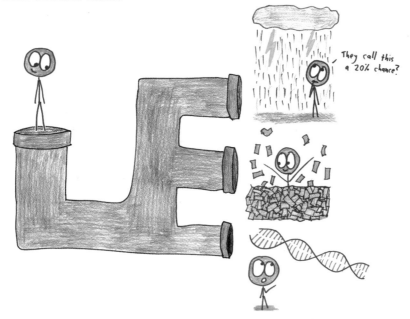

If this description rings false to you, it's perhaps because you've been to a place called "school." If so, you have my condolences.

When I graduated from college in 2009, I thought I knew why mathematics was unpopular: It was, on the whole, badly taught. Math class took a beautiful, imaginative, logical art, shredded it into a bowl of confetti, and assigned students the impossible, mind-numbing task of piecing the original back together. No wonder they groaned. No wonder they failed. No wonder adults look back on their math experiences with a shudder and a gag reflex. The solution struck me as obvious: Math required better explanations, better explainers.

Then I became a teacher. Undertrained and oozing hubris, I needed a brutal first year in the classroom to teach me that, whatever I knew about math, I didn't yet understand math education, or what the subject meant to my students.

One day that September, I found myself leading an awkward impromptu discussion of why we study geometry. Did grown-ups write two-column proofs? Did engineers work in "no calculator" environments? Did personal finance demand heavy use of the rhombus? None of the standard justifications rang true. In the end, my 9th graders settled on "We study math to prove to colleges and employers that we are smart and hardworking." In this formulation, the math itself didn't matter. Doing math was a weightlifting stunt, a pointless show of intellectual strength, a protracted exercise in résumé building. This depressed me, but it satisfied them, which depressed me even more.

The students weren't wrong. Education has a competitive zero-sum aspect, in which math functions as a sorting mechanism. What they were missing—what I was failing to show them—was math's deeper function.

Why does mathematics underlie everything in life? How does it manage to link disconnected realms—coins and genes, dice and stocks, books and baseball? The reason is that mathematics is a system of thinking, and every problem in the world benefits from thinking.

Since 2013, I've been writing about math and education—sometimes for publications like *Slate*, the *Atlantic*, and the *Los Angeles Times*, but mostly for my own blog, *Math with Bad Drawings*. People still ask me why I do the bad drawings. I find this odd. No one ever wonders why I choose to cook mediocre food, as if I've got a killer chicken l'orange that, on principle, I decline to serve. It's the same with my art. *Math with Bad Drawings* is a less pathetic title than *Math with the Best Drawings I Can Manage; Honestly, Guys, I'm Trying*, but in my case they are equivalent.

I suppose my path began one day when I drew a dog on the board to illustrate a problem, and got the biggest laugh of my career. The students found my ineptitude shocking, hilarious, and, in the end, kind of charming. Math too often feels like a high-stakes competition; to see the alleged expert reveal himself as the worst in the room at something—anything—that can humanize him and, perhaps, by extension, the subject. My own humiliation has since become a key element in my pedagogy; you won't find that in any teacher training program, but, hey, it works.

Lots of days in the classroom, I strike out. Math feels to my students like a musty basement where meaningless symbols shuffle back and forth. The kids shrug, learn the choreography, and dance the tuneless dance.

But on livelier days, they see distant points of light and realize that the basement is a secret tunnel, linking everything they know to just about everything else. The students grapple and innovate, drawing connections, taking leaps, building that elusive virtue called "understanding."

Unlike in the classroom, this book will sidestep the technical details. You'll find few equations on these pages, and the spookiest ones are decorative anyway. (The hard core can seek elaboration in the endnotes.) Instead, I want to focus on what I see as the true heart of mathematics: the concepts. Each section of this book will tour a variety of landscapes, all sharing the underground network of a single big idea: How the rules of geometry constrain our design choices. How the methods of probability tap the liquor of eternity. How tiny increments yield quantum jumps. How statistics make legible the mad sprawl of reality.

Writing this book has brought me to places I didn't expect. I hope that reading it does the same for you.

—BEN ORLIN, October 2017

$$(1-\lambda)^2 \pi \sqrt{a^2+b^2} \qquad |u \wedge v| = |u||v|\sin\theta$$

$$b = \underline{b} - \underline{a} = \lambda(c - a) \qquad (\underline{u} \cdot \underline{v})^2 + |\underline{u} \wedge \underline{v}|^2 = |\underline{u}|^2|\underline{v}|^2$$

$$\underline{c} \qquad \lambda = \frac{1}{2} \qquad |\underline{u} \wedge \underline{v}|^2 = |\underline{u}|^2|\underline{v}|^2 \sin^2\theta$$

$$\underline{u} \wedge \underline{v} = |\underline{u}||\underline{v}|\sin\theta \, \underline{n}$$

$$|AB| : |BC| = \lambda : (1-\lambda)^2 \pi \sqrt{a^2+b^2} \qquad |u \wedge v| = |u||v|\sin\theta$$

$$\vec{AC} = (\underline{c} - \underline{a}), \vec{AB} = \underline{b} - \underline{a} = \lambda(c - a) \qquad (\underline{u} \cdot \underline{v})^2 + |\underline{u} \wedge \underline{v}|^2 = |u|$$

$$b = 11 - \lambda]g + \lambda \underline{c} \qquad \lambda = \frac{1}{2} \qquad |\underline{u} \wedge \underline{v}|^2 = |u|^2|v|^2 s_i$$

$$b = \frac{1}{2}(\underline{a} + \underline{c}) \qquad \underline{u} \wedge \underline{v} = |\underline{u}||\underline{v}|\sin\theta$$

$$\underline{u} \wedge \underline{v} = A\underline{n}$$
$$A = |u||v|\sin\theta$$
$$t + \delta t$$
$$r(t+\delta t)$$
$$\delta r$$
$$r(t)$$
$$\delta r = r(t+\delta t) - r(t)$$
$$dt \to 0$$

$$\underline{d} = \frac{1}{2}(b+c), \qquad \underline{g} = \frac{1}{3}(a+b+c)$$

$$\underline{g} = \frac{1}{3}(a+b+c)$$

area A
$$|u||v|\sin\theta$$

$$\frac{dr}{dt} = \lim_{\delta t \to 0} \frac{\delta r}{\delta t}$$

$$\wedge v|^2 = |\underline{u}|^2|\underline{v}|^2$$

$$(\underline{u} \cdot \underline{v})^2 + |\underline{u} \wedge \underline{v}|^2 = |\underline{u}|^2|\underline{v}|^2$$

$$\mu_3 U_3)^2 + (U_2 U_3 - U_5 U_2)^2 -$$
$$(U_4 U_2 - U_2 U_4) =$$

$$(U_1 U_1 + U_2 U_2 + U_3 U_3)^2 + (U_2 U_3 - U_5 U_2)^2 -$$
$$(U_1 V_1 - U_4 U_3)^2 + (U_4 V_2 - U_2 U_4)^2 =$$

$$\frac{2}{3})(U_1^2 + U_2^2 \frac{2}{3}) \qquad (\theta, \varphi) \in [0, 2\pi] \times \qquad = (u_1^2 + u_2^2 + u_3^2)(U_1^2 + U_2^2 \frac{2}{3})$$

$$x^2 + y^2 = z^2 \qquad [-\pi/2, \pi/2] \qquad x^2 + y^2 = z^2$$

$$(1-\lambda)^2 \pi \sqrt{a^2+b^2} \qquad |u \wedge v| = |u||v|\sin\theta$$

$$\vec{B} = \underline{b} - \underline{a} = \lambda(c - a) \qquad (\underline{u} \cdot \underline{v})^2 + |\underline{u} \wedge \underline{v}|^2 = |\underline{u}|^2|\underline{v}|^2$$

$$\lambda_c \qquad \lambda = \frac{1}{2} \qquad |\underline{u} \wedge \underline{v}|^2 = |\underline{u}|^2|v|^2 \sin^2\theta$$

$$\underline{u} \wedge \underline{v} = |\underline{u}||\underline{v}|\sin\theta \, \underline{n}$$

$$|AB| : |BC| = \lambda : (1-\lambda)^2 \pi \sqrt{a^2+b^2} \qquad |u \wedge v| = |u||v|\sin\theta$$

$$\vec{AC} = (\underline{c} - \underline{a}), \vec{AB} = \underline{b} - \underline{a} = \lambda(c - a) \qquad (\underline{u} \cdot \underline{v})^2 + |\underline{u} \wedge \underline{v}|^2 = |$$

$$b = 11 - \lambda]g + \lambda \underline{c} \qquad \lambda = \frac{1}{2} \qquad |\underline{u} \wedge \underline{v}|^2 = |u|^2|v|^2 s$$

$$b = \frac{1}{2}(\underline{a} + \underline{c}) \qquad \underline{u} \wedge \underline{v} = |\underline{u}||\underline{v}|\sin\theta$$

$$\underline{u} \wedge \underline{v} = A\underline{n}$$
$$A = |u||v|\sin\theta$$
$$t + \delta t$$
$$r(t+\delta t)$$
$$\delta r$$
$$r(t)$$
$$\delta r = r(t+\delta t) - r(t)$$
$$dt \to 0$$

$$), \quad \underline{g} = \frac{1}{3}(a+b+c) \qquad \underline{d} = \frac{1}{2}(b+c), \qquad \underline{g} = \frac{1}{3}(a+b+c)$$

area A
$$|u||v|\sin\theta$$

$$\frac{dr}{dt} = \lim_{\delta t \to 0} \frac{\delta r}{\delta t}$$

$$\wedge v|^2 = |\underline{u}|^2|\underline{v}|^2$$

$$(\underline{u} \cdot \underline{v})^2 + |\underline{u} \wedge \underline{v}|^2 = |\underline{u}|^2|\underline{v}|^2$$

$$l_3 U_3)^2 + (U_2 U_3 - U_5 U_2)^2 -$$
$$(U_4 U_2 - U_2 U_4) =$$

$$(U_1 U_1 + U_2 U_2 + U_3 U_3)^2 + (U_2 U_3 - U_5 U_2)^2 -$$
$$(U_1 V_1 - U_4 U_3)^2 + (U_4 V_2 - U_2 U_4)^2 =$$

$$u_3^2)(U_1^2 + U_2^2 \frac{2}{3}) \qquad (\theta, \varphi) \in [0, 2\pi] \times \qquad = (u_1^2 + u_2^2 + u_3^2)(U_1^2 + U_2^2 \frac{2}{3})$$

$$x^2 + y^2 = z^2 \qquad [-\pi/2, \pi/2] \qquad x^2 + y^2 = z^2$$

$$(1-\lambda)^2 \pi \sqrt{a^2+b^2} \qquad |u \wedge v| = |u||v|\sin\theta$$

$$\vec{B} = \underline{b} - \underline{a} = \lambda(c - a) \qquad \vec{AC} = (\underline{c} - \underline{a}), \vec{AB} = \underline{b} - \underline{a} = \lambda(c - a) \qquad (\underline{u} \cdot \underline{v})^2 + |\underline{u} \wedge \underline{v}|^2$$

I

HOW TO THINK LIKE A MATHEMATICIAN

To be honest, mathematicians don't do much. They drink coffee, frown at chalkboards. Drink tea, frown at students' exams. Drink beer, frown at proofs they wrote last year and can't for the life of them understand anymore.

It's a life of drinking, frowning, and, most of all, thinking.

You see, there are no physical objects in math: no chemicals to titrate, no particles to accelerate, no financial markets to destroy. Rather, the verbs of the mathematician all boil down to actions of thought. When we calculate, we turn one abstraction into another. When we give proofs, we build logical bridges between related ideas. When we write algorithms or computer programs, we enlist an electronic brain to think the thoughts that our meat brains are too slow or too busy to think for themselves.

Every year that I spend in the company of mathematics, I learn new styles of thought, new ways to use that nifty all-purpose tool inside the skull: How to master a game by fussing with its rules. How to save thoughts for later, by recording them in loopy Greek symbols. How to learn from my errors as if they were trusted professors. And how to stay resilient when the dragon of confusion comes nibbling at my toes.

In all these ways, mathematics is an action of the mind.

What about math's vaunted "real-world" usefulness? How can spaceships and smartphones and godforsaken "targeted ads" emerge from this fantasy skyline of pure thought? Ah, patience, my friend. All that comes later. We've got to begin where all mathematics begins: That is to say, with a game . . .

Chapter 1

ULTIMATE TIC-TAC-TOE

WHAT IS MATHEMATICS?

Once at a picnic in Berkeley, I saw a group of mathematicians abandon their Frisbees to crowd around the last game I would have expected: tic-tac-toe.

As you may have discovered yourself, tic-tac-toe is terminally dull. (That's a medical term.) Because there are so few possible moves, experienced players soon memorize the optimal strategy. Here's how all my games play out:

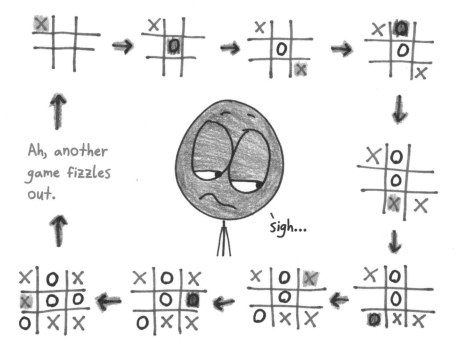

Ah, another game fizzles out.

`sigh...

If both players have mastered the rules, every game will end in a draw, forever and ever. The result is rote gameplay, with no room for creativity.

But at that picnic in Berkeley, the mathematicians weren't playing ordinary tic-tac-toe. Their board looked like this, with each square of the original nine turned into a mini-board of its own:

As I watched, the basic rules became clear:

1. Each turn, you mark a square on a mini-board.

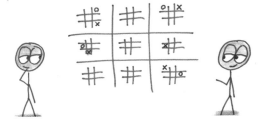

2. When you get three in a row on a mini-board, you win that board.

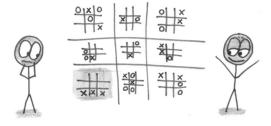

3. When you get three mini-boards in a row, you win the game.

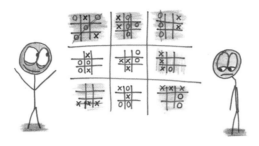

But it took a while longer for the most important rule of the game to emerge:

You don't get to pick which of the nine mini-boards to play on. Instead, that's determined by your opponent's previous move. **Whichever square she picks on the mini-board, you must play next in the corresponding square on the big board.**

(And whichever square *you* pick on that mini-board will determine which mini-board *she* plays on next.)

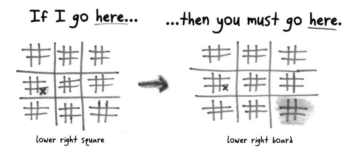

If I go <u>here</u>...   ...then you must go <u>here</u>.

lower right square   lower right board

And if you go <u>here</u>...   ...then I must go <u>here</u>.

top center square   top center board

This lends the game its strategic element. You can't just focus on each mini-board in turn. You've got to consider where your move will send your opponent, and where her next move will send you—and so on, and so on.

(There's only one exception to this rule: If your opponent sends you to a mini-board that's already been won, then congratulations—you can go anywhere you like, on any of the other mini-boards.)

The resulting scenarios look bizarre, with players missing easy two- and three-in-a-rows. It's like watching a basketball star bypass an open layup and fling the ball into the crowd. But there's a method to the madness. They're thinking ahead, wary of setting up their foe on prime real estate. A clever attack on a mini-board leaves you vulnerable on the big board, and vice versa—it's a tension woven into the game's fabric.

From time to time, I play Ultimate Tic-Tac-Toe with my students; they enjoy the strategy, the chance to defeat a teacher, and, most of all, the absence of trigonometric functions. But every so often, one of them will ask a sheepish and natural question: "So, I like the game," they'll say, "but what does any of this have to do with math?"

I know how the world sees my chosen occupation: a dreary tyranny of inflexible rules and formulaic procedures, no more tension-filled than, say, insurance enrollment, or filling out your taxes. Here's the sort of tedious task we associate with mathematics:

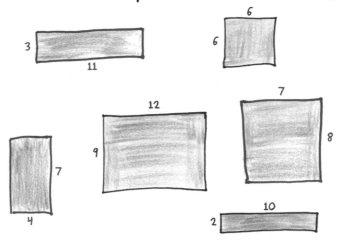

Find the area and perimeter of each rectangle.

This problem can hold your attention for a few minutes, I guess, but before long the actual concepts slip from mind. "Perimeter" no longer means the length of a stroll around the rectangle's border; it's just "the two numbers, doubled and added." "Area" doesn't refer to the number of one-by-one squares it takes to cover the rectangle; it's just "the two numbers, multiplied." As if competing in a game of garden-variety tic-tac-toe, you fall into mindless computation. There's no creativity, no challenge.

But, as with the revved-up game of Ultimate Tic-Tac-Toe, math has far greater potential than this busywork suggests. Math can be bold and exploratory, demanding a balance of patience and risk. Just trade in the rote problem above for one like this:

Create two rectangles so that the first has a bigger perimeter, and the second a bigger area.

This problem embodies a natural tension, pitting two notions of size (area and perimeter) against one another. More than just applying a formula, solving it requires deeper insight into the nature of rectangles. (See the endnotes for spoilers.)

Or, how about this:

Create two rectangles so that the first has exactly twice the perimeter of the second, and the second has exactly twice the area of the first.

A little spicy, right?

In two quick steps, we've traveled from sleepwalking drudgery to a rather formidable little puzzle, one that stumped a whole school's worth of bright-eyed 6th graders when I tossed it as a bonus on their final exam. (Again, see the endnotes for solutions.)

As this progression suggests, creativity requires freedom, but freedom alone is not enough. The pseudopuzzle "Draw two rectangles" provides loads of freedom but no more spark than a damp match. To spur real creativity, a puzzle needs constraints.

Take Ultimate Tic-Tac-Toe. Each turn, you've got only a few moves to choose from—perhaps three or four. That's enough to get your imagination flowing, but not so many that you're left drowning in a sea of incalculable possibilities. The game supplies just enough rules, just enough constraints, to spur our ingenuity.

That pretty well summarizes the pleasure of mathematics: creativity born from constraints. If regular tic-tac-toe is math as people know it, then Ultimate Tic-Tac-Toe is math as it ought to be.

Math as Most People Know It

Math as It Should Be

You can make the case that *all* creative endeavors are about pushing against constraints. In the words of physicist Richard Feynman, "Creativity is imagination in a straitjacket." Take the sonnet, whose tight formal restrictions—*Follow this rhythm! Adhere to this length! Make sure these words rhyme! Okay . . . now express your love, lil' Shakespeare!*—don't undercut the artistry but heighten it. Or look at sports. Humans strain to achieve goals (*kick the ball in the net*) while obeying rigid limitations (*don't use your hands*). In the process, they create bicycle kicks and diving headers. If you ditch the rulebook, you lose the grace. Even the wacky, avant-garde, convention-defying arts—experimental film, expressionist painting, professional wrestling—draw their power from playing against the limitations of the chosen medium.

Creativity is what happens when a mind encounters an obstacle. It's the human process of finding a way through, over, around, or beneath. No obstacle, no creativity.

But mathematics takes this concept one step further. In math, we don't just follow rules. We invent them. We tweak them. We propose a possible constraint, play out its logical consequences, and then, if that way leads to oblivion—or worse, to boredom—we seek a new and more fruitful path.

For example, what happens if I challenge one little assumption about parallel lines?

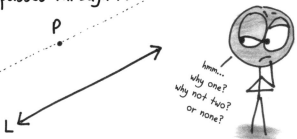

There is exactly one line that is parallel to L and passes through P.

P

L

hmm...
why one?
why not two?
or none?

Euclid laid out this rule regarding parallel lines in 300 BCE; he took it for granted, calling it a fundamental assumption (a "postulate"). This struck his successors as a bit funny. Do you really have to assume it? Shouldn't it be provable? For two millennia, scholars poked and prodded at the rule, like a piece of food caught between their teeth. At last they realized: Oh! It *is* an assumption. You can assume otherwise. And if you do, traditional geometry collapses, revealing strange alternative geometries, where the words "parallel" and "line" come to mean something else entirely.

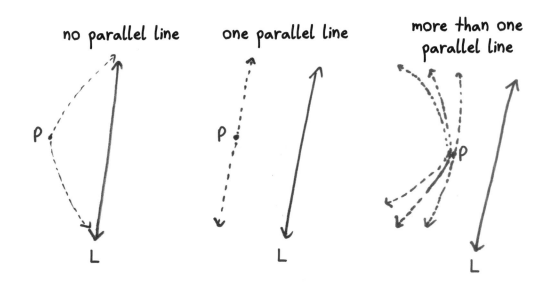

no parallel line

one parallel line

more than one parallel line

P

L

P

L

P

L

New rule, new game.

As it turns out, the same holds true of Ultimate Tic-Tac-Toe. Soon after I began sharing the game around, I learned that there's a single technicality upon which everything hinges. It comes down to a question I confined to a parenthetical earlier: What happens if my opponent sends me to a mini-board that's already been won?

These days, my answer is the one I gave above: Since that mini-board is already "closed," you can go wherever you want.

But originally, I had a different answer: As long as there's an empty space on that mini-board, you have to go there—even though it's a wasted move.

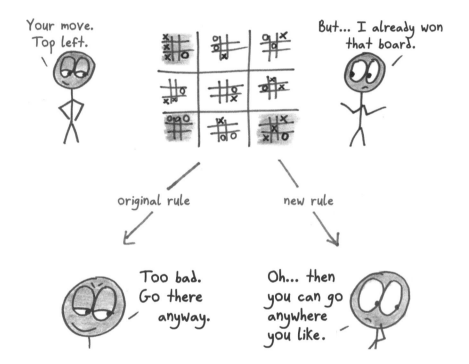

This sounds minor—just a single thread in the tapestry of the game. But watch how it all unravels when we give this thread a pull.

I illustrated the nature of the original rule with an opening strategy that I dubbed (in a triumph of modesty) "The Orlin Gambit." It goes like this:

Take the very center square.

O will go somewhere else.

Take the center again.

O will grab two in a row.

Take the center again.

O wins the board, laughing.

Take the center again.

O begins to see your trickery...

Take the center again.

And again. And again.

When all is said and done, X has sacrificed the center mini-board in exchange for superior position on the other eight. I considered this stratagem pretty clever until readers pointed out its profound uncleverness. The Orlin Gambit wasn't just a minor advantage. It could be extended into a guaranteed winning strategy. Rather than sacrifice one mini-board, you can sacrifice two, giving you two-in-a-rows on each of the other seven. From there, you can ensure victory in just a handful of moves.

Embarrassed, I updated my explanation to give the rule in its current version—a small but crucial tweak that restored Ultimate Tic-Tac-Toe to health.

New rule, new game.

This is exactly how mathematics proceeds. You throw down some rules and begin to play. When the game grows stale, you change it. You pose new constraints. You relax old ones. Each tweak creates new puzzles, fresh challenges. Most mathematics is less about solving someone else's riddles than about devising your own, exploring which constraints produce interesting games and which produce drab ones. Eventually, this process of rule tweaking, of moving from game to game, comes to feel like a grand never-ending game in itself.

Mathematics is the logic game of inventing logic games.

The history of mathematics is the unfolding of this story, time and time again. Logic games are invented, solved, and reinvented. For example: what happens if I tweak this simple equation, changing the exponents from 2 into another number, like 3, or 5, or 797?

<u>Equation</u>

<u>New Equations</u>

$$a^2 + b^2 = c^2 \longrightarrow$$

$$a^3 + b^3 = c^3$$
$$a^5 + b^5 = c^5$$
$$a^{797} + b^{797} = c^{797}$$

etc.

Oh! I've turned an ancient and elementary formula, easily satisfied by plugging in whole numbers (such as 3, 4, and 5), into perhaps the most vexing equation that humans have ever encountered: Fermat's last theorem. It tormented scholars for 350 years, until the 1990s, when a clever Brit locked himself in an attic and emerged nearly a decade later, blinking at the sunlight, with a proof showing that whole number solutions can never work.

Or, what happens if I take two variables—say, x and y—and create a grid that lets me see how they relate?

$x = 0, \quad y = -4$

$x = 1, \quad y = -3$

$x = 2, \quad y = 0$

$x = 3, \quad y = 5$

$x = -1, \quad y = -3$

$x = -2, \quad y = 0$

$x = -3, \quad y = 5$

$x = \frac{1}{2}, \quad y = -3\frac{3}{4}$

$x = -\frac{3}{2}, \quad y = \frac{-7}{4}$

too much data!

ah, so clear!

Oh! I've invented graphing, and thereby revolutionized the visualization of mathematical ideas, because my name is Descartes and that's why they pay me the big bucks.

Or, consider that squaring a number always gives a positive answer. Well, what if we invented an exception: a number that, when squared, is *negative*? What happens then?

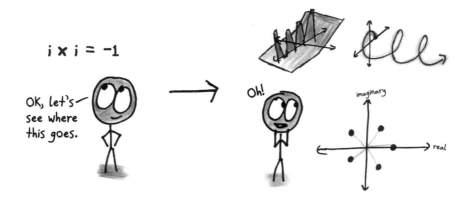

Oh! We've discovered imaginary numbers, thereby enabling the exploration of electromagnetism and unlocking a mathematical truth called "the fundamental theorem of algebra," all of which will sound pretty good when we add it to our résumés.

In each of these cases, mathematicians at first underestimated the transformative power of the rule change. In the first case, Fermat thought the theorem would yield to a simple proof; as centuries of his frustrated successors will attest, it didn't. In the second case, Descartes's graphing idea (now called "the Cartesian plane" in his honor) was originally confined to the appendix of a philosophical text; it was often dropped from later printings. And in the third case, imaginary numbers faced centuries of snubs and insults ("as subtle as they are useless," said the great Italian mathematician Cardano) before being embraced as true and useful numbers. Even their name is pejorative, originating from a dis first uttered by none other than Descartes.

It's easy to underestimate new ideas when they emerge not from somber reflection but from a kind of play. Who would guess that a little tweak to the rules (new exponent; new visualization; new number) could transform a game beyond imagination and recognition?

I don't think the mathematicians at that picnic had this in mind as they huddled over a game of Ultimate Tic-Tac-Toe. But they didn't need to. Whether we're cognizant of it or not, the logic game of inventing logic games exerts a pull on us all.

Chapter 2

WHAT DOES MATH LOOK LIKE TO STUDENTS?

Alas, this will be a short, bleak chapter. I'd apologize for that, but I'll be too busy apologizing for other things, like the often soul-pureeing experience of math education.

You know what I'm talking about. To many students, "doing math" means executing the prescribed sequence of pencil maneuvers. Mathematical symbols don't symbolize; they just dance across the page in baffling choreography. Math is a tale told by an abacus, full of "sin" and "θ," signifying nothing.

$$\frac{\frac{7x-1}{2}+4}{7}-3=8$$

A train's position is described by यनर्हृ . You are tied to the tracks and wriggle out of your rope at a variable rate given by गुरूशाऽपि If mistaking संस्कृ and तम्सं means instant death, how long until आनंद हू? (Assume no air resistance.)

Allow me to offer two brief apologies:

First, I apologize to my own students for any times I've made you feel this way about math. I've tried not to; then again, I've also tried to stay on top of email, cut back on ice cream, and avoid four-month lapses between haircuts. Please enter my plea as "merely human."

Second, I apologize to mathematics, for any violence you have suffered at my hands. In my defense, you're an intangible tower of quantitative concepts bound together by abstract logic, so I doubt I've left any lasting scars. But I'm not too proud to say I'm sorry.

That's it for this chapter. I promise the next one will have more explosions, like any good sequel should.

Chapter 3

WHAT DOES MATH LOOK LIKE TO MATHEMATICIANS?

It's very simple. Math looks like language.

A funny language, I'll admit. It's dense, terse, and painstaking to read. While I zip through five chapters of a *Twilight* novel, you might not even turn the page in your math textbook. This language is well suited to telling certain stories (e.g., the relations between curves and equations), and ill suited to others (e.g., the relations between girls and vampires). As such, it's got a peculiar lexicon, full of words that no other tongue includes. For example, even if I could translate $a_0 + \sum_{n=1}^{\infty} \left(a_n \cos \frac{n\pi x}{L} + b_n \sin \frac{n\pi x}{L} \right)$ into plain English, it wouldn't make sense to someone unfamiliar with Fourier analysis, any more than *Twilight* would make sense to someone unfamiliar with teenage hormones.

But math is an ordinary language in at least one way. To achieve comprehension, mathematicians employ strategies familiar to most readers. They form mental images. They paraphrase in their heads. They skim past distracting technicalities. They draw connections between what they're reading and what they already know. And—strange as it may seem—they engage their emotions, finding pleasure, humor, and squeamish discomfort in their reading material.

Now, this brief chapter can't teach fluent math any more than it could teach fluent Russian. And just as literary scholars might debate a couplet by Gerard Manley Hopkins or the ambiguous phrasing of an email, so mathematicians will disagree on specifics. Each brings a unique perspective, shaped by a lifetime of experience and associations.

That said, I hope to offer a few nonliteral translations, a few glimpses into the strategies by which a mathematician might read some actual mathematics. Consider it Squiggle Theory 101.

When mathematicians see "7 × 11 × 13"...

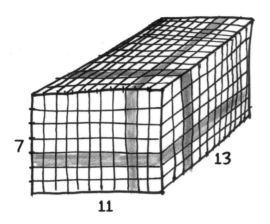

A common question I get from students: "Does it matter whether I multiply by 11 or 13 first?" The answer ("no") is less interesting than what the question reveals: that in my students' eyes, multiplication is an *action*, a thing you *do*. So one of the hardest lessons I teach them is this: Sometimes, *don't*.

You don't have to read 7 × 11 × 13 as a command. You can just call it a number and leave it be.

Every number has lots of aliases and stage names. You could also call this number 1002 – 1, or 499 × 2 + 3, or 5005/5, or Jessica, the Number That Will Save Planet Earth, or plain old 1001. But if 1001 is how this number is known to its friends, then 7 × 11 × 13 isn't some quirky and arbitrary moniker. Rather, it's the official name you'd find on the birth certificate.

7 × 11 × 13 is the prime factorization, and it speaks volumes.

Some key background knowledge: Addition is kind of boring. To wit, writing 1001 as the sum of two numbers is a truly dull pastime: you can do it as 1000 + 1, or 999 + 2, or 998 + 3, or 997 + 4 . . . and so on, and so on, until you slip into a boredom coma. These decompositions don't tell us anything special about 1001, because all numbers can be broken up in pretty much the same way. (For example, 18 can be written as 17 + 1, or 16 + 2, or 15 + 3 . . .) Visually, this is like breaking a number up into two piles. No offense, but piles are dumb.

Multiplication: now that's where the party's at. And to join the festivities, you need to deploy our first math-reading strategy: **forming mental images**.

As the picture on the previous page shows, multiplication is all about grids and arrays. 1001 can be seen as a giant structure of blocks, 7 by 11 by 13. But that's just getting started.

You can visualize this as 13 layers of 77 each. Or, if you tilt your head sideways, it's 11 layers of 91 each. Or, tilting your head a different sideways, seven layers of 143 each. All of these ways to decompose 1001 are immediately evident from the prime factorization . . . but virtually impossible to discern from the name 1001 without laborious guesswork.

The prime factorization is the DNA of a number. From it, you can read all the factors and factorizations, the numbers that divide our original and the numbers that don't. If math is cooking class, then $7 \times 11 \times 13$ isn't the pancake recipe. It's the pancake itself.

When mathematicians see "$A = \pi r^2$"...

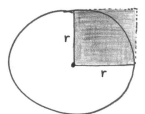

"To fill the circle, you'd need π squares."
a little more than 3

To casual fans, π is a mystic rune, a symbol of mathematical sorcery. They ponder its irrationality, memorize thousands of its digits, and commemorate Pi Day on March 14 by combining the most glorious of mankind's arts (dessert pies) with the least glorious (puns). To the general public, π is an object of obsession, awe, even something approaching worship.

And to mathematicians, it's roughly 3.

That infinite spool of decimal places that so captivates laypeople? Well, mathematicians aren't that bothered. They know math is about more than precision; it's about quick estimates and smart approximations. When building intuition, it helps to streamline and simplify. **Intelligent imprecision** is our next crucial math-reading strategy.

Take the formula A = πr², which many students have heard so often the mere phrase "area of a circle" triggers them to scream, "Pi *r* squared!" like sleeper agents programmed by brainwashing. What does it mean? Why is it true?

Well, forget the 3.14159. Let your mind go fuzzy. Just look at the shapes.

r is the radius of our circle. It's a length.

*r*², then, is the area of a little square, like the one pictured.

Now, the π-dollar question: How does the area of the *circle* compare to the area of the *square*?

Clearly, the circle is bigger. But it's not quite four times bigger (since four squares would cover the circle and then some). Eyeballing it, you might speculate that the circle is a little more than three times bigger than the square.

And that's exactly what our formula says: *Area = a little more than 3 × r²*.

If you want to verify the precise value—why 3.14-ish and not 3.19-ish?—then you can use a proof. (There are several lovely demonstrations; my favorite involves peeling the circle like an onion and stacking the layers to make a triangle.) But mathematicians, whatever they insist, don't always prove everything from first principles. Like everyone from carpenters to zookeepers, they're happy to employ a tool without knowing precisely how it was constructed, so long as they have a sense of why it works.

When mathematicians see "$y = \frac{1}{x^2}$"...

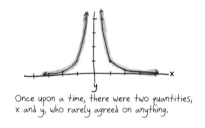

Once upon a time, there were two quantities, x and y, who rarely agreed on anything.

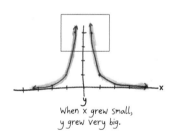

When x grew small, y grew very big.

When x grew big, y grew very, very small.

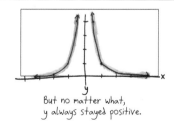

But no matter what, y always stayed positive.

"Go graph these equations" is a familiar homework assignment. I've given it myself. It's also the seed of a vicious myth: that graphs are an end unto themselves. In fact, graphing is not like solving an equation or carrying out an operation. A graph is not an end; it is always, *always* a means.

A graph is a tool for data visualization, a picture that tells a story. It represents another powerful math-reading strategy: **turning the static into the dynamic**.

Take the equation above: $y = \frac{1}{x^2}$. Here, x and y stand for a pair of numbers obeying a precise relationship. Here are some examples of possible pairs:

x	2	3	4	5
y	$\frac{1}{4}$	$\frac{1}{9}$	$\frac{1}{16}$	$\frac{1}{25}$

A few patterns are already peeking out. But our vision is only as powerful as our technology, and tables are not a fancy gadget. Of the infinite x-y pairs that satisfy this equation, the table can show only a handful at a time, like the scrolling text of a stock ticker. We need a superior visualization tool: the mathematical equivalent of a television screen.

Enter the graph.

By treating x and y as a sort of longitude and latitude, we transform each intangible pair of numbers into something satisfyingly geometric: a point. The infinite set of points becomes a unified curve. And from this, a story emerges, a tale of motion and change.

- When x grows small, approaching zero ($\frac{1}{5}$, $\frac{1}{60}$, $\frac{1}{1000}$ . . .), then y balloons to enormous values (25, 3600, 1 million . . .).
- When x grows large (20, 40, 500 . . .), then y shrinks to tiny proportions ($\frac{1}{400}$, $\frac{1}{16,000}$, $\frac{1}{250,000}$ . . .).
- While x can assume negative values (-2, -5, -10), y never does. It remains positive.
- And neither variable ever adopts a value of zero.

Okay, maybe it's not the juiciest plotline, but mental maneuvers like this mark the difference between the novice's experience of mathematics (as a paralyzing stream of meaningless symbols) and the mathematician's (as something coherent and communicative). Graphs imbue lifeless equations with a sense of motion.

When mathematicians see "(x−5)(x−7) = 0"...

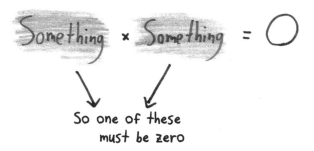

So one of these
must be zero

There's a psychological phenomenon known by the unfortunate name of **chunking.** More than just a way to cleanse the system after too many beers, chunking also happens to be a potent mental technique, indispensable to mathematicians. It's our next math-reading strategy.

When we "chunk," we reinterpret a set of scattered, hard-to-retain details as a single unit. The above equation offers a simple example. A good chunker disregards the minutiae on the left side. Is it x or y, 5 or 6, + or -? Don't know; don't care. Instead, you see only a pair of chunky factors, forming a skeletal equation that reads: $chunk \times chunk = 0$.

Now, if you're familiar with multiplication, you know that zero is a peculiar result to come by.

6×5? Not zero.

18×307? Not zero.

$13.91632 \times 4{,}600{,}000{,}000{,}000$? No need to fire up your calculator app: This, too, is not zero.

Zero is a singular number in the world of multiplication. In contrast to, say, 6, which can be produced a variety of ways (3×2, 1.5×4, $1200 \times 0.005 \ldots$), zero is special and elusive. In fact, there's precisely one way that two numbers multiply to give you zero: if one of the original numbers was, itself, zero.

This is where our chunking pays off: Since their product is zero, one of the chunks must be, too. If it was the first one (x - 5), then x must be 5. And if it was the latter (x - 7), then x must be 7.

Equation solved.

Chunking purifies not just the contents of our stomachs but the contents of our minds. It makes the world digestible. And the more you learn, the more aggressively you are able to chunk. A high schooler might chunk a whole row of algebra as "find the area of the trapezoid." An undergraduate might chunk several dense lines of calculus as "compute the volume of the solid of revolution." And a graduate student might chunk half a page of formidable Greek-lettered technicalities as "compute the Hausdorff dimension of the set." For each level up, you've got to learn subtle new details: What are trapezoids? How do integrals behave? What was Hausdorff smoking, and where can we get some?

But we don't learn details for details' sake. We learn details so as to ignore them later, to focus on the big, chunky picture instead.

When mathematicians see "x^2 vs. 2^x"...

vs.

Switch two symbols. What happens?

Well, in the eyes of the novice, nothing. You've interchanged scribbles, swapped syllables in a gibberish language. Who cares? But in the eyes of a mathematician, it can be like switching sea and sky, or mountain and cloud, or bird and fish (much to the consternation of both). Switching two symbols can change *everything*.

For example, take the two expressions above and imagine that x is 10.

Now, 10^2 is a big number. It means 10×10. That's 100. It's a reasonable number of students to teach in a given year, or miles to drive to a theme park, or dollars to pay for a used television. (It's a suspicious number of dalmatians to own.)

But 2^{10} is a much *bigger* number. It's $2 \times 2 \times 2 \times 2 \times 2 \times 2 \times 2 \times 2 \times 2 \times 2$. That's 1024. It's a reasonable number of students to teach in a decade, or miles to drive to the world's greatest theme park, or dollars to pay for an awe-inspiring television. (It's a *very* suspicious number of dalmatians to own; this is why we have animal cruelty laws.)

As we try out bigger x's, the gap between the two expressions widens. In fact, the word "widens" is too soft, like describing the Grand Canyon as "bit of a crack in the ground." As x grows, the gap between x^2 and 2^x *explodes*.

To wit, 100^2 is quite big. It's 100×100. That's 10,000.

But 2^{100} is *huge*. It's $2 \times 2 \times 2 \times 2 \times 2 \times 2 \times 2 \times 2 \times 2 \times 2 \times 2 \times 2 \times 2 \times 2$ $\times 2 \times 2 \times 2 \times 2 \times 2 \times 2 \times 2 \times 2 \times 2 \times 2 \times 2 \times 2 \times 2 \times 2 \times 2 \times 2 \times 2 \times 2$ $\times 2 \times 2 \times 2 \times 2 \times 2 \times 2 \times 2 \times 2 \times 2 \times 2 \times 2 \times 2 \times 2 \times 2 \times 2 \times 2 \times 2 \times 2$ $\times 2 \times 2 \times 2 \times 2 \times 2 \times 2 \times 2 \times 2 \times 2 \times 2 \times 2 \times 2 \times 2 \times 2 \times 2 \times 2 \times 2 \times 2$ $\times 2 \times 2 \times 2 \times 2 \times 2 \times 2 \times 2 \times 2 \times 2 \times 2 \times 2 \times 2 \times 2 \times 2 \times 2 \times 2 \times 2 \times 2$ $\times 2 \times 2 \times 2 \times 2 \times 2 \times 2$. That's 1,267,650,600,228,229,401,496,703,205,376, which is roughly a billion billion trillion.

If we're talking about pounds, then the former is the weight of a pickup truck carrying a load of bricks. Heavy, to be sure, but the latter is in a different weight class altogether.

It's the size of a hundred thousand Earths.

x^2 and 2^x don't look so different to the untrained eye. But the more math you experience and the more fluent you become in this language of squiggles, the more dramatic the difference begins to feel. Before long it becomes visceral, tactile; it starts to **enlist your emotions**, which is our final crucial strategy. You read lines of mathematics with a full spectrum of feelings, from satisfaction to sympathy to shock.

Eventually, mixing up x^2 and 2^x becomes as absurd as imagining a pickup truck towing a hundred thousand planets behind it.

Chapter 4

HOW SCIENCE AND MATH SEE EACH OTHER

1. TWINS NO MORE

Back in 9th grade, my friend John and I looked oddly similar: pensive round-faced brown-haired boys who spoke about as often as furniture. Teachers called us by the wrong names; older students thought we were one person; and in the yearbook, we're mislabeled as each other. For all I remember, we sparked a friendship just to mess with people.

Then, with time, we thinned out and grew up. John is now a broad-chested six foot two and looks like a Disney prince. I'm five nine and have been described as "a cross between Harry Potter and Daniel Radcliffe." The "practically twins" stage of our friendship is long gone.

And as it is with me and John, so it is with mathematics and science.

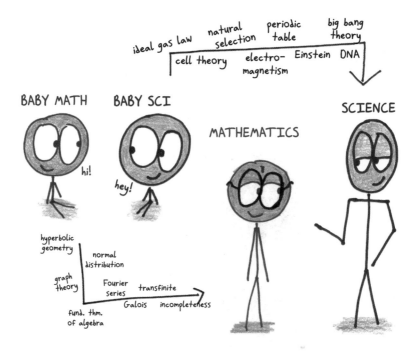

Back when science and math still had their baby fat, they didn't just look the same. They *were* the same. Isaac Newton didn't worry whether history would classify him as a scientist or a mathematician: he was both, inextricably. The same went for his intellectual elder siblings—Galileo, Kepler, Copernicus. For them, science and math were interwoven, inseparable. Their core insight was that the physical cosmos followed mathematical recipes. Objects obeyed equations. You couldn't study one without the other, any more than you could eat a single ingredient out of a baked cake.

Science and math have diverged since then. Just look at how they're taught: separate classrooms, separate teachers, separate (if equally mind-numbing) textbooks. They've thinned out, added muscle, aged beyond their wide-eyed 9th-grade innocence.

But still, people get them confused. Any fool can see that I'm not John, and John isn't me, but the citizen on the street has more trouble telling the differences between math and science—especially at a layman's distance.

Perhaps the easiest way to tell them apart is to answer this question: What do science and math look like, not to the layperson, but to each other?

2. IN EACH OTHER'S EYES

Looking through science's eyes, the answer is pretty clear. Science sees mathematics as a tool kit. If science is a golfer, then mathematics is the caddy, proffering the right club for each situation.

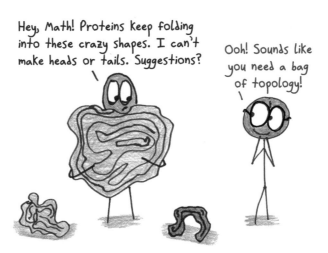

This view casts mathematics in a subservient role. It's not my favorite dynamic, but, hey, I sympathize. Science is trying to make sense of reality, which—as you'll know if you have experience with reality—is pretty darn hard. Things are born. Things die. Things leave maddeningly partial fossil records. Things exhibit qualitatively different behaviors at quantum and relativistic scales. Reality is kind of a mess.

Science quests to make sense of all this. It aims to predict, classify, and explain. And in this endeavor, it sees mathematics as a vital assistant: the Q to its James Bond, supplying helpful gadgets for the next adventure.

Now, let's spin the camera 180° and change perspectives. How does mathematics see science?

You'll find that we're not just alternating camera angles. We're changing film genres altogether. Whereas science sees itself as the protagonist of an action movie, mathematics sees itself as the auteur director of an experimental art project.

That's because, on a fundamental level, mathematicians do not care about reality.

I'm not talking about the odd habits of mathematicians—muttering to themselves, wearing the same trousers for weeks, forgetting their spouse's names on occasion. I'm talking about their work. Despite the aggressive ad campaign about its "real-world usefulness," mathematics is pretty indifferent to the physical universe.

What math cares about are not *things* but *ideas.*

Math posits rules and then unpacks their implications by careful reasoning. Who cares if the resulting conclusions—about infinitely long cones and 42-dimensional sausages—don't resemble physical reality? What matters is their abstract truth. Math lives not in the material universe of science but in the conceptual universe of logic.

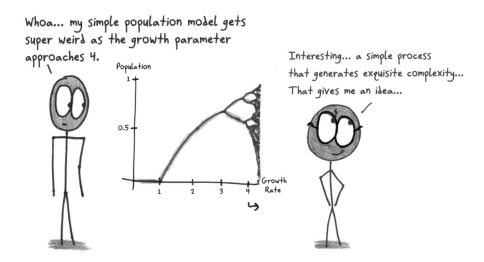

Mathematicians call this work "creative." They liken it to art.

That makes science their muse. Think of a composer who hears chirping birds and weaves the melody into her next work. Or a painter who gazes at cumulus clouds drifting through an afternoon sky, and models her next landscape on that image. These artists don't care if they've captured their subjects with photorealistic fidelity. For them, reality is nothing more or less than a fertile source of inspiration.

That's how math sees the world, too. Reality is a lovely starting point, but the coolest destinations lie far beyond it.

3. THE PARADOX OF MATHEMATICS

Math sees itself as a dreamy poet. Science sees it as a supplier of specialized technical equipment. And herein we find one of the great paradoxes of human inquiry: These two views, both valid, are hard to reconcile. If math is an equipment supplier, why is its equipment so strangely poetic? And if math is a poet, then why is its poetry so unexpectedly useful?

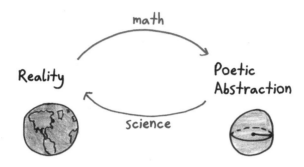

To see what I mean, take the twisted history of knot theory.

This branch of mathematics, like many, was inspired by a scientific problem. Before the discovery of atoms, some scientists (including Lord Kelvin) entertained the idea that the universe was filled with a substance called ether, and matter was made of knots and tangles in it. Thus, they sought to classify all the possible knots, creating a periodic table of tangles.

Before long, science lost interest, lured away by the shiny new theory of atoms (which had the unfair advantage of being right). But mathematics was hooked. It turns out that classifying knots is a delightful and devilish problem. Two versions of the same knot can look wildly different. Totally different knots can taunt you with their resemblance. It was perfect fuel for mathematicians, who soon developed an exquisite, complex theory of knots, unperturbed that their clever abstractions appeared to have no practical purpose whatsoever.

The centuries rolled along.

Then, science ran into a real snake of a problem. As you know, every cell inscribes its precious information on DNA molecules, which are fantastically long. If straightened out and laid flat, the DNA from one of your cells would stretch for six feet—a hundred thousand times the length of the cell itself. This makes DNA a long string stuffed into a small container. If you've ever shoved earbuds into your pocket or removed Christmas lights from their box, you know what this scenario creates: maddening tangles. How do bacteria manage this? Can we learn their tricks? Can we perhaps disable cancer cells by tangling their DNA?

Biology was flummoxed. It needed help. "Ooh," Mathematics cried. "I know just the thing!"

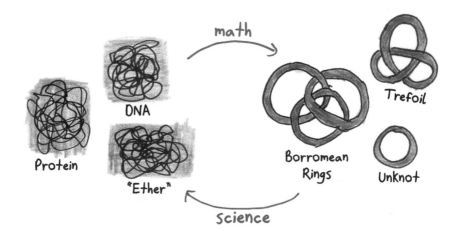

Here, then, is a brief biography of knot theory. It was born from a practical need. Soon, it grew into something deliberately impractical, a logic game for poets and philosophers. And yet somehow this mature creature, which had barely spared a thought for reality over the years, became profoundly useful in a field far removed from the one of its birth.

This is no isolated case. It's a basic pattern of mathematical history.

Remember the strange alternate geometries from Chapter 1? For centuries, scholars saw them as mere novelties, a poet's fancy. They bore no correspondence to reality, which was understood to follow Euclid's assumption about parallel lines.

Then, along came a young patent clerk named Einstein. He realized that these wacko alternatives aren't just thought experiments; they underlie the structure of the cosmos. From our tiny perspective, the universe appears Euclidean, much in the way that the curved Earth appears flat. But zoom out, shedding the prejudices of the surface dweller, and you'll see a different picture entirely, a shifting landscape of strange curvatures.

Those "useless" geometries turned out to be pretty darn useful.

Perhaps my favorite example concerns logic itself. Early philosophers like Aristotle developed symbolic logic ("If p, then q") as a guide for scientific thinking. Then mathematical theorists got their hands on it, and turned logic into something bizarre and abstract. Reality fell away. By the 20th century, you had folks like Bertrand Russell writing Latin-titled tomes that aimed to "prove" from elementary assumptions that $1 + 1 = 2$. What could be more useless, more irredeemable?

One logician's mother nagged him: *C'mon, honey, what's the point of all this abstract mathematics? Why not do something useful?*

That mother was named Ethel Turing. And as it turns out, her son Alan was kind of on to something: a logic machine we now call "the computer."

I can't blame her for doubting. Who would have guessed that her son's abstract research into logical systems would help to define the next century? No matter how many examples I encounter, this historical cycle of useful to useless to useful again remains a wonder and a mystery to me.

My favorite description of this phenomenon is a phrase coined by physicist Eugene Wigner: "the unreasonable effectiveness of mathematics." After all, bacteria don't know any knot theory, so why should they follow its rules? The space-time continuum hasn't studied hyperbolic geometry, so why does it execute its theorems so perfectly? I've read some philosophers' answers to this question, but I find them tentative and conflicting, and none have worked to blunt my astonishment.

So, how best to understand the relationship between the poet we call Math and the adventurer known as Science? Perhaps we ought to see them as a symbiotic pair of very different creatures, like an insect-eating bird perched on the back of a rhino. The rhino gets its itchy problems solved. The bird gets nourished. Both emerge happy.

When you visualize math, picture something dainty and elegant astride the wrinkled gray mass of reality below.

Chapter 5

GOOD MATHEMATICIAN VS. GREAT MATHEMATICIAN

It's a lot of fun to bust myths. Just look at the carefree explosions and ear-to-ear smiles of television's MythBusters, and you can see it's a career with high job satisfaction.

What's trickier is tweaking myths. Lots of the culture's prevailing views about mathematics aren't flat-out wrong—just crooked, or incomplete, or overemphasized. Is computation important? Sure, but it's not all-important. Does math require attention to detail? Yes, but so do knitting and parkour. Was Carl Gauss a natural genius? Well, yeah, but most beautiful math comes not from depressive German perfectionists but from ordinary folks like you and me.

Before we close this section, this chapter offers one final exploration of how to think like a mathematician, a chance to revise and annotate some popular myths. Like most myths, they've got a basis in truth. And, like most myths, they miss the flux, uncertainty, and struggle toward understanding that make us human—and that make us mathematicians.

A few years ago, when I lived in England, I taught a boy named Corey. He reminded me of a soft-spoken 12-year-old Benjamin Franklin: quiet and insightful, with long ginger hair and round spectacles. I can totally picture him inventing bifocals.

Corey poured his heart into every homework assignment, drew lucid connections across topics, and packed up his papers at the period's end with such care and patience that I always fretted he'd be late for his next lesson. So it's no surprise that on the first big test in November, Corey nailed every question.

Well . . . every question that he'd had time to answer.

The bell rang with the last quarter of his test still blank. He scored in the low 70s and came to me the next day with a furrowed brow. "Sir," he said, because England is an amazing land where clumsy 29-year-old teachers get fancy honorifics, "why are tests timed?"

I figure honesty is the best policy. "It's not because speed is so important. We just want to see what students can do by themselves, without anyone's help."

"So why not let us keep working?"

"Well, if I held the class hostage for a whole day, it might annoy your other teachers. They want you to know about science and geography, because of their nostalgic attachment to reality."

I realized that I had never seen Corey like this: jaw clenched, eyes dark. He was radiating frustration. "I could have answered more," he said. "I just ran out of time."

I nodded. "I know."

There wasn't much else to say.

Intentionally or not, school mathematics sends a loud, clear message: *Speed is everything.* Tests are timed. Early finishers get to start their homework. Just look how periods end: with a ringing bell, as if you've just finished a round in a perverse, compulsory, logarithm-themed game show. Math comes to feel like a race, and success becomes synonymous with quickness.

All of which is supremely silly.

Speed has one fabulous advantage: It saves time. Beyond that, mathematics is about deep insight, real understanding, and elegant approaches, none of which you're likely to find when moving at 600 miles per hour. You learn more mathematics by thinking carefully than by thinking fast, just as you learn more botany by studying a blade of grass than by sprinting like the dickens through a wheat field.

Corey understands this. I only hope that teachers like me don't manage, against our own best intentions, to persuade him otherwise.

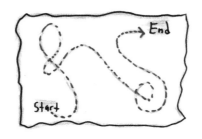

A good mathematician has the patience to reach complicated answers.

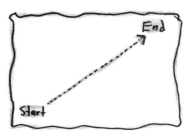

A great mathematician has the patience to reach simple answers.

My wife, who is a research mathematician, once pointed me toward a funny pattern in mathematical life.

- *Step #1*: There's a tricky and exciting question on the loose, an important conjecture in need of proof. Many try to tame the beast, without success.
- *Step #2*: Someone finally proves it, via a long and convoluted argument that's full of insight but very difficult to follow.
- *Step #3*: Over time, new proofs are published, growing shorter and simpler, shorter and simpler, until eventually the original proof is relegated to "historical" status: an inefficient Edison-era lightbulb made obsolete by sleeker, more modern designs.

Why is this trajectory so common?

Well, the first time you arrive at the truth, you've often come by a jagged and meandering path. To your credit, it takes patience to survive all the twists and turns. But it takes a more profound patience to keep thinking afterward. Only then can you sort the necessary steps from the superfluous ones, and perhaps boil that sprawling 120-page proof down to a seamless 10-pager.

A good mathematician can remember all the details.

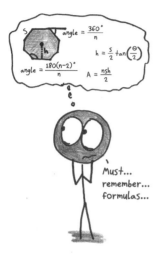

Must... remember... formulas...

A great mathematician can forget all the details.

Eh, just break everything into triangles.

In 1920, of all the branches of mathematics, algebra was perhaps the dullest. To do algebra was to enter a bog of petty specifics, a thornbush of unwieldy technicalities, a discipline of details.

Then, in 1921, a mathematician named Emmy Noether published a paper titled "Theory of Ideals in Ring Domains." It was the dawn, a colleague later said, of "abstract algebra as a conscious discipline." Noether wasn't interested in unpacking particular numerical patterns. In fact, she laid the whole idea of "number" to the side. All that mattered to her were symmetry and structure. "She taught us to think in simple, and thus general, terms," remembered another colleague years later. "She therefore opened a path to the discovery of algebraic regularities where before these regularities had been obscured."

To do good work, you've first got to engage with nitty-gritty details. Then, to do great work, you've got to move beyond them.

For Noether, abstraction was more than an intellectual habit; it was a way of life. "She often lapsed into her native German when she was bothered by some idea," said a colleague. "She loved to walk. She would take her students off for a jaunt on a Saturday afternoon. On these trips, she would become so absorbed in her conversation on mathematics that she would forget about the traffic, and her students would need to protect her."

Great mathematicians don't bother with trivialities like crosswalks and traffic flow. They've got their mind's eye on something bigger.

A good mathematician tackles the problem head-on.

A great mathematician circles around it.

In 1998, Sylvia Serfaty got hooked on a question. It dealt with how certain vortices evolve through time. She had literally written the book on the subject—*Vortices in the Magnetic Ginzburg-Landau Model*—but found herself stumped by this puzzle.

"A lot of good research," she later said, "actually starts from very simple things, elementary facts, basic bricks...Progress in math comes from understanding the model case, the simplest instance in which you encounter the problem. And often it is an easy computation; it's just that no one had thought of looking at it this way."

You can attack the castle at the main gate, fighting the defensive forces head-on. Or, you can seek to better understand the castle itself—and perhaps uncover an easier way in.

The mathematician Alexander Grothendieck offered a different metaphor for this matter of perspective. Think of a problem as a scrumptious hazelnut, with "nourishing flesh protected by the shell." How do we get at it?

There are two basic approaches. First, the hammer and chisel: whack the shell with a sharp edge until you've cracked it. That's likely to work, but it's violent and effortful. Or, second, you can immerse the nut in water.

As Grothendieck put it, "From time to time, you rub so the liquid penetrates better, and otherwise you let time pass. The shell becomes more flexible through weeks and months—when the time is ripe, hand pressure is enough; the shell opens like a perfectly ripened avocado!"

For two decades, progress came to Serfaty in fits and starts, as she and collaborators let the nut soak. Finally, in 2015, she found the right angle of attack, the perfect perspective. The problem yielded in a matter of months.

A good mathematician selects the most powerful tool for the job.

Welding torch it is!

A great mathematician selects the least powerful tool for the job.

Nothing a little duct tape can't fix.

Every mathematical field has its Holy Grail. For many statisticians, that grail was the Gaussian correlation inequality.

"I know of people who worked on it for 40 years," said a Penn State statistician named Donald Richards. "I myself worked on it for 30." Many scholars gave valiant attempts—hundred-page calculations, sophisticated geometric framings, new developments drawn from analysis and probability theory—but none could capture the grail. A few came to suspect it might even be false, a myth.

Then, one day in 2014, Richards received an email from a German retiree named Thomas Royen. It contained a Microsoft Word attachment. That was weird—virtually all serious mathematicians type up their work with a program called LaTeX. Why was this former pharmaceutical employee reaching out to a leading researcher in statistics?

Well, as it so happens, this fellow had proved the Gaussian correlation inequality. He had done it using arguments and formulas that any graduate student could follow. The insight had popped into his head while he was brushing his teeth.

"When I looked at it," said Richards, "I knew instantly that it was solved." Richards felt humbled and frustrated to have missed such a simple argument, but he was delighted nevertheless. "I remember thinking to myself that I was glad to have seen it before I died. Really, I was so glad I saw it."

Royen's tale echoes a piece of wisdom I learned from a beloved physics teacher: "If you want to kill flies, you don't need bazookas." For mathematicians, there is elegance in restraint.

November 2010, Oakland, California. I am 23 years old and spending the morning demonstrating to my trigonometry students the sheer number of ways that one can unsuccessfully explain de Moivre's theorem.

"Okay, let me start again!" I say, sweating. "You're going to be raising this number to the nth power anyway, right? And so you can go ahead and add $2k\pi/n$ to the angle, because the zoombini's other fraggle will repatriate the fleen. You get it now?"

"No!" a roomful of teenagers cried, hands over their ears. "Stop! You're making it worse!"

A student named Vianney raised her hand. "Can I try to see if I understand?"

"Please," I sighed. "Be my guest."

"Okay, we're going to be dividing this angle in half, right?"

The room calmed a little.

"And adding 360° to the angle just puts you back in the same place, right? Like, 90° and 450° are no different; you've just done an extra rotation."

Students straightened in their chairs.

"But when you divide the angle in half, that extra 360° becomes an extra 180°. You wind up on the opposite side of the circle."

I watched as mental lightbulbs flashed all over the room, like a crowd of photographers.

"So," Vianney concluded, "that's why there are two solutions. Right?"

I thought for a moment. The whole room leaned forward. At last I nodded. "Yes. Well said."

The room erupted into applause, as an avalanche of high fives and shoulder thumps buried Vianney. Even I could see the superiority of her approach. Where I had been trying to explain the theorem backward, and to cover all values of n simultaneously, Vianney turned the problem right side out and focused on the case where $n = 2$.

History's greatest mathematicians have left their mark not only by feats of intellectual strength but by blazing trails and teaching others to follow. Euclid consolidated past insights into an irreplaceable textbook. Cantor distilled his fresh understanding of infinity into clean, easy-to-follow arguments. Stein mentored generations of harmonic analysts, becoming adviser to mathematicians as great as himself.

It's not that Vianney understood de Moivre's theorem better than I did. It's that she could cash out her knowledge into clear language, whereas my insight remained locked inside a thick skull, attached to a clumsy tongue. A mathematician who can't convey his thinking will suffer the same fate that I did that day: to be a lonely island of thought, whose ideas never reach other shores, while the mathematician who can share her truth enjoys a hero's welcome from the grateful crowd.

A good mathematician
wants to be the best.

A great mathematician
wants to learn from
the best.

No! I can't lose!

Ooh! Look at that technique!

I doubt you've heard of Ngô Bảu Châu, unless (a) You got bored one day and decided to memorize every winner of the Fields Medal, math's most prestigious prize, or (b) You're from Vietnam, where Châu is a national celebrity and household name.

(By the way, huge kudos to Vietnam for that. America's closest approximation to a celebrity mathematician is Will Hunting: not even our most famous Matt Damon character.)

Châu is a fierce competitor, who craved as a youngster to be the best of the best. For a time, he was. Winning back-to-back gold medals at the International Mathematical Olympiad, he became the pride of his school and the envy of his peers, the Simone Biles of Vietnamese math.

But his undergraduate years brought a slow quicksand agony, as he gradually realized that he didn't understand the math he was learning. "My professors thought I was a fantastic student," he remembers. "I could do all the exercises. I did well on exams. But I didn't understand anything." Achievement came to feel like hollow orb: a brittle shell of accolades, soon to shatter and reveal the horrible vacuum inside.

Then came a turning point. He stopped trying to be the best, and began to learn from them.

He shovels credit on his PhD adviser Gérard Laumon. "I had one of the best advisers in the world," Châu says, glowing. "I would come to his office every week. He would read with me one or two pages every time." They went line by line, equation by equation, settling for nothing less than full comprehension.

Châu soon began work on the famous Langlands program. Think of it as the transcontinental railroad of modern mathematics: a sweeping vision for how to connect several distant branches of the discipline. The project has drawn generations of ambitious mathematicians like Châu into its orbit, and Châu found himself attracted to a particularly vexing piece of the Langlands program: proving the "fundamental lemma."

Was it the Olympiad all over again? Rival mathematicians jockeying for primacy, racing to be the first to prove it?

No, says Châu.

"I was helped a lot by people in my field," he says. "Many people encouraged me to do so, in a very sincere way. I asked them for advice, and they would tell me what to learn. It was very open. I did not feel competition." With help from these collaborators, Châu managed to prove the fundamental lemma. It was the work than won him his Fields.

For such an exceptional scholar, Châu's tale is pleasantly ordinary. People who thrive in the tournament atmosphere of school—with its clear rankings, easy lateral comparisons, and steady diet of rewards—find they need a new attitude as they advance into the open-ended world of scholarship. Drawn in as competitors, they evolve into collaborators.

$(1-\lambda)^2 \pi \sqrt{a^2+b^2}$

$|\underline{u} \wedge \underline{v}| = |\underline{u}||\underline{v}|\sin\theta$

$|\overline{ABl} : |\overline{BCl}| = \lambda : (1-\lambda)^2 \pi \sqrt{a^2+b^2}$

$|\underline{u} \wedge \underline{v}| = |\underline{u}||\underline{v}|\sin\theta$

$\underline{b} = \underline{b} - \underline{a} = \lambda(c-a)$

$(\underline{u} \cdot \underline{v})^2 + |\underline{u} \wedge \underline{v}|^2 = |\underline{u}|^2|\underline{v}|^2$

$\overrightarrow{AC} = (c-a), \overrightarrow{AB} = \underline{b} - \underline{a} = \lambda(c-a)$

$(\underline{u} \cdot \underline{v})^2 + |\underline{u} \wedge \underline{v}|^2 =$

$\underline{c} \quad \lambda = \frac{1}{2}$

$|\underline{u} \wedge \underline{v}|^2 = |\underline{u}|^2|\underline{v}|^2 \sin^2\theta$

$b = 11 - \lambda 1 g + \lambda \underline{c} \quad \lambda = \frac{1}{2}$

$|\underline{u} \wedge \underline{v}|^2 = |\underline{u}|^2|\underline{v}|^2$

$\underline{u} \wedge \underline{v} = |\underline{u}||\underline{v}|\sin\theta \underline{n} \quad 2\pi\sqrt{a^2+b^2} \quad \underline{b} = \frac{1}{2}(\underline{y}+\underline{c})$

$\underline{u} \wedge \underline{v} = |\underline{u}||\underline{v}|\sin\theta$

$\underline{u} \wedge \underline{v} = A\underline{n}$

$A = |\underline{u}||\underline{v}|\sin\theta$

$t + \delta t \quad r(t+\delta t)$

$\underline{d} = \frac{1}{2}(b+c), \quad \underline{g} = \frac{1}{3}(a+b+c)$

$), \quad \underline{g} = \frac{1}{3}(a+\underline{b}+\underline{c})$

area A $\quad |\underline{u}||\underline{v}|\sin\theta$

δr

$r(t)$

$\delta \underline{r} = \underline{r}(t+\delta t) - \underline{r}(t)$

$dt \to 0$

$\frac{d\underline{r}}{dt} = \lim_{\delta t \to 0} \frac{\delta \underline{r}}{\delta t}$

$\wedge \underline{v}|^2 = |\underline{u}|^2|\underline{v}|^2$

$(\underline{u} \cdot \underline{v})^2 + |\underline{u} \wedge \underline{v}|^2 = |\underline{u}|^2|\underline{v}|^2$

$_5 u_3)^2 + (u_2 u_3 - u_5 u_2)^2 -$

$(u_1 u_1 + u_2 u_2 + u_5 u_3)^2 + (u_2 u_3 - u_5 u_2)^2 -$

$(u_4 v_2 - u_2 v_1) =$

$(u_5 v_1 - u_4 u_3)^2 + (u_4 v_2 - u_2 v_1) =$

$\frac{2}{3})(U_1^2 + U_2^2 + \frac{2}{3})$

$x^2 + y^2 = z^2$

$(\theta, \varphi) \in [0, 2\pi] \times$

$= (u_1^2 + u_2^2 + u_3^2)(U_1^2 + U_2^2 + \frac{2}{3})$

$x^2 + y^2 = z^2$

$[-\pi/2, \pi/2]$

II

DESIGN
THE GEOMETRY OF STUFF THAT WORKS

H ere's a tough-love life lesson: just because you set your mind to something doesn't mean you can do it.

Say you're creating a square, and you want its diagonal to be the same length as its sides. Well, I hate to break it to you, but that's not how squares work.

Okay ... do that again, except <u>square</u> !

Nope. Never.

Or say you're tiling a floor, and you want to use a pattern of regular pentagons. I'm sorry to report that this will never pan out. There will always be gaps between your tiles.

Insolent pentagons! Why do you defy me?

We tile for _no_ one.

Nothing personal, dude.

Just not our thing.

Or say you're building an equilateral triangle, and you fancy a bespoke angle size—say 70°, or 49°, or 123°. I'm afraid the cosmos doesn't care: Those angles are going to be 60°, no more and no less, or else your triangle won't be equilateral.

Why so uncooperative, angle? Be more like your siblings!

Trust me: the angles know what they're doing.

Human laws are flexible, subject to repeal and renegotiation. The drinking age is 21 in the US, 18 in Britain, 16 in Cuba, "never" in Afghanistan, and "step right up" in Cambodia. Any country can stiffen or relax its drinking laws at a whim (and of course, the stiffer your drink and the laxer your laws, the more whim-prone you become). Geometry's laws are not like that. There's no wiggle room: no president to issue pardons, no jury to acquit, no officer to let you off with a warning. Math's rules are self-enforcing, unbreakable by their nature.

Yet as we've seen, and shall see many times again, that's not a bad thing. Restrictions birth creativity. The laws on what shapes *can't* do come packaged with case studies illuminating what they *can*. In design projects ranging from sturdy buildings to useful paper to planet-destroying space stations, geometry inspires even as it constrains.

So forget the sentiment that "anything is possible!" It is very sweet but deeply unnatural, like most things we feed to children. The reality is harsher—and more wondrous.

Chapter 6

WE BUILT THIS CITY ON TRIANGLES

I'd like you to meet this chapter's star: the triangle.

Aw, shucks!

It's not your typical protagonist. Snooty literary types may dismiss it as two-dimensional. Yet this atypical hero will embark on a typical hero's journey: rising from humble origins, learning to harness an inner strength, and ultimately serving the world in a time of crisis.

Now, if your mind is so narrow that it cannot encompass the idea of a valiant polygon, then by all means, read no further. Don your blindfold of prejudice. Just make sure to squeeze your eyelids tight, for only the deepest darkness can guard your shuttered thoughts from the radiant truth, the penetrating light, of planar geometry. Don't you remember? We built this city. We built this city on triangles . . .

1. TWELVE KNOTS IN AN EGYPTIAN ROPE

Welcome to ancient Egypt: a prosperous kingdom, a teeming bureaucracy, rigid of faith and right-angled of elbow. It shall endure for millennia as the sun rises and sets on empires with lesser headgear.

Come, let's stroll. The year is 2570 BCE, and the Great Pyramid of Giza is halfway built. Three and a half million tons of brick rise out of the desert, with 3 million more to follow. The heaviest blocks outweigh two bull elephants. The square base is 756 feet per side, the length of three New York City blocks. Already, it is the tallest structure in the world. Upon its completion a decade from now, when its workforce of 80,000 can finally rest for lemonade, it will stand 481 feet tall. Five millennia from today, it will still be here, history's most enduring skyscraper, the greatest triumph of triangular architecture.

Except it isn't.

I mean, it is still standing, last I checked. But it is not a victory for triangles. If you want to see the triangle at work, forget the Great Pyramid, and walk with me to a vacant lot nearby. There, we find a small team of surveyors carrying a peculiar loop of rope, in which they have tied 12 equally spaced knots.

What for? Just watch. After a few steps, three men each grab a knot (#1, #4, and #8, respectively) and pull the rope tight. As if by magic, it forms a right-angled triangle. A fourth worker marks out the right angle. They let the triangle relax once again into a knotty loop. The scene repeats again and again, until the whole expanse is subdivided into perfect sections of equal size.

If you've ever stayed awake through a geometry lesson (and even if you haven't), the scene may call to mind Pythagoras's theorem. This rule states that if you draw a square on each side of a right-angled triangle, the two smaller areas taken together will equal the larger one. Or, in modern algebraic terms: $a^2 + b^2 = c^2$.

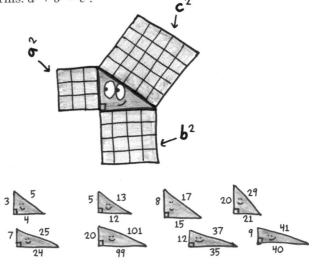

There is an endless roster of such triangles. For example, the sides can have lengths 5, 12, and 13; or 7, 24, and 25; or 8, 15, and 17; or my personal favorite, 20, 99, and 101. The Egyptians wisely chose the simplest example, the triangle with side lengths 3, 4, and 5. Hence, the 12 knots.

But this chapter is not about Pythagoras and "his" rule (which earlier civilizations found without his help). It is about a simpler and more fundamental property of triangles, a hidden elegance we shall soon discover. The story of the triangle begins not in a Pythagorean temple, nor atop the Great Pyramid, but here, in a vacant field. A slack rope transforms into a surveyor's tool. It's our first hint of a power so great it will make the pyramids look crude by comparison.

2. THREE PLANKS, ONE SELF

Now this tale enters its psychological phase. The triangle must peer into its own soul and ask the essential question: *Who am I?*

Am I a shape like any other, indistinguishable except by the tally of my sides and the number of my corners? The music swells, and the triangle begs the cosmos for a sign, a vision, a purpose. *What,* it cries, *am I truly made of?* Then it comes, a thunderous voice deep from within.

I am made of three sides.

Okay, maybe not so revelatory, as moments of self-discovery go. It's a bit like a therapy patient noticing the couch. But there are hidden depths here. New truths emerge when we see triangles not as holistic shapes but as assembled three-part composites.

For example: to make a triangle, not just any three sides will work. Take the lengths 10 centimeters, 3 centimeters, and 2 centimeters. These three segments form only a gap-toothed nonshape—an untriangle, if you will. The long side is too long; the short ones are too short. I dub it a "T. rex triangle," because the stubby arms can't reach across the body.

Well, I'm insulted, but I see the resemblance.

3 cm   2 cm   10 cm

It's a universal truth: a triangle's longest side must be shorter than the other two put together.

This law is self-evident to a fly walking the perimeter of the triangle. It knows that a straight path (from A to B) will always be shorter than a detour (from A to C to B). Thus, the two shorter sides, taken together, must be longer than the third side.

This law has a companion, even deeper and more powerful: **If three sides *do* form a triangle, then they will form *exactly one*.** Given three sides, there is no room to embellish or improvise. There is a single mold to follow.

For example, let's agree in advance upon three sides (such as 5 meters, 6 meters, and 7 meters), then build our personal triangles in separate rooms. I guarantee that we will emerge with identical creations.

Watch: I'll lay my longest side flat along the ground, then lean the other two against each other to meet at their tips. Done! Slide the corner to the left, and one side sticks out; to the right, and the other does. The solution is what mathematicians call *unique*. Even without anticipating your method, I know you'll arrive at the same solution, because there is no other.

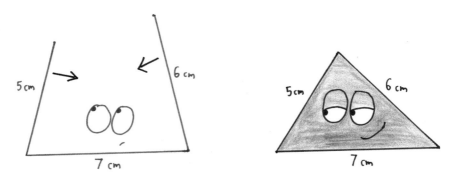

This truth belongs to the triangle alone. No other polygon can claim the same.

Try it with the triangle's closest cousin, the four-sided quadrilateral. I lay one side flat. I stand up the next two vertically. And I lay my final side across the top, fastening the corners with duct tape for good measure. But then, a wind begins to blow. My square sways. The uprights tilt, hinging at the corners, and the whole apparatus starts to collapse like a folding chair. Every moment brings a new shape, from "square" to "almost square" to "kind of diamondy" to "skinny rhombus of super pointiness."

Wait... I'm not special?

The four sides haven't yielded a unique shape. Rather, they've given us an infinite family of possibilities. Any member can be rearranged into another by applying a little pressure.

Thus, we see the triangle's hidden magic, its secret identity: not its mere three-sided-ness, but the rigidity it confers.

The Egyptian rope pullers knew this power. By pulling their 12-knot rope tight, they summoned a Pythagorean triangle into existence, conjuring a right angle from rope. You might try instead to summon a square, with its four right angles, but beware: a whole family of unwanted shapes will answer that call. Even pulled tight, the quadrilateral will twist at its corners, shifting forms, refusing definition. The same is true of pentagons, hexagons, heptagons, and all the other polygon cousins. None can do what the triangle can.

The pyramids, being solid, did not take advantage of this special power. Cubes, cones, frustums—any of these would have suited the pharaohs' purpose just as well. The blunt language of stone doesn't much care what shape it articulates.

Now, I don't mean to slight the pyramids. For one, "slight" is not the right word for a brick pile weighing 20 billion pounds. I admire their otherworldly precision: sides equal to within 8 inches, facing the cardinal directions with an error below 0.1°, with corners less than 0.01° from right-angled perfection. Those Egyptian cats knew their math.

But I must point out that these are a surveyor's triumphs, not an engineer's. The Great Pyramid remains, fundamentally, a big stack of blocks. That's cool for a towering symbol of your pharaoh's immortality, but not great for a building you're actually hoping to, y'know, use. The pyramid's modest chambers and slivers of passageway occupy less than 0.1% of its internal volume. Imagine if the Empire State Building were solid steel except for a single floor with a 2-foot ceiling, and you too will begin aspiring to a more efficient building plan.

In centuries to come, architects would seek new poetries of structure. They would build bridges wider than the sky, towers taller than Babel. And for this, they would need a shape of extraordinary resilience, a shape of unique and unyielding character: a three-sided, three-cornered hero.

3. THE BENDING RAFTERS
OF THE BURDENED WORLD

At this point, our story intersects another—the 10-millennium saga of human architecture. A quick recap of what you've missed:

1. "Outside" is a bad place to live. It can get very cold, there's nowhere to keep your stuff, and sometimes bears show up. That's why humans invented "inside."
2. To create "inside," you make a big empty shape and then live in the middle of it.
3. If your shape is friendly and made of the right things, then it will be nice to live inside and won't fall down on you. This is called "architecture."

Okay, now that we're up to speed, I can introduce you to a crucial supporting character in the triangle's story: the beam. If you're an architect aiming to avoid both (a) pyramid-like monoliths and (b) collapsing floors, then beams will likely factor in your design.

The beam's effect is to take vertical forces and move them horizontally. For example, picture a plank stretched over a ditch. When you stand on the beam, your weight pushes it downward. But the real support isn't below: it's off to the sides, where the plank meets the dirt. The beam takes a force in the middle and moves its impact to the sides.

There's just one problem: beams are inefficient.

Architecture, like life itself, is all about stress management. Whereas life offers many types of stresses (deadlines, child-rearing, low phone battery, etc.), structures experience just two: pushes and pulls. Pushes create *compression*, when an object is squeezed tighter. Pulls create *tension*, when an object is stretched apart. Each type has its own peculiar character, and different materials cope in different ways. Concrete can withstand fantastic degrees of compression, but it crumbles under tension. At the other extreme, steel cables can endure incredible amounts of tension, but they buckle under the slightest compression.

Now, picture a beam sagging under a load. It curves into a smile (or, more fitting, a grimace). What's the nature of its strain: tension or compression?

The answer is "both." Look at the upper surface of the beam: like a runner on the inside lane of a track, it has a shorter distance to curve. Thus, its material squeezes together, creating compression. Now, move your gaze to the bottom: like a runner on the outside lane of a track, it has *farther* to curve, so its material stretches and experiences tension.

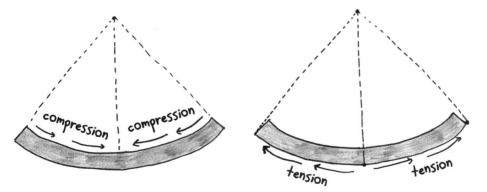

So far, nothing to worry about: many materials, such as wood, cope well with both pushes and pulls. The problem isn't that the beam experiences both stresses; it's that a large fraction of the beam experiences *neither*.

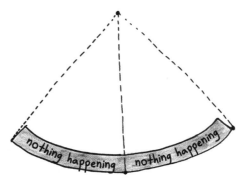

Look to the center. Halfway between the compression on top and the tension on bottom, the beam's middle experiences no strain whatsoever. Its curve is the carefree smile of a guy who's not helping. The middle is wasted substance, no better than the useless bulk of a pyramid. A generic beam squanders half of its strength, like a student coasting on 50% effort.

Every teacher knows the next two words: That's unacceptable. In architecture, every ounce counts, whether you're building a sky-tickling tower, a canyon-spanning bridge, or a life-affirming roller coaster.

Rest assured: architects are no fools. They've got a plan.

4. THE SHAPE OF THE RESISTANCE

Did I say architects are no fools? I may have to walk that back once you hear their solution. Since the beam's top and bottom take all of the strain while the middle freeloads on their efforts, the brilliant solution devised by architects was to build beams *without middles*.

No need to shout; I get it. A beam with no middle is otherwise known as "two separate beams" and isn't much of a solution.

Unless . . . you keep just a *little bit* of the middle. While you move most of the substance to the extremes, you leave a thin connective layer in between. The resulting shape's cross-section resembles a capital letter *I*; hence, the term "I-beam."

It is I.
I - Beam.

This is a good start. But we've still got wasted material in the center. Thus, we activate phase two of the architects' plan: start punching holes in the I-beam.

Every gap in the material saves precious resources while costing us almost nothing in strength. The emptier the space, the greater our savings, which means our best hope is to leave the I-beam's middle a hole-riddled web, more vacant space than solid substance.

Uh... I - Beam cool with this.
Mostly.

But hang on. Before we start punching holes willy-nilly, we need a plan. What patterns of gaps will minimize the material required while preserving the structure's strength and rigidity? Where can we turn for a design that's simple and resilient, not to mention well suited to the flat, nearly two-dimensional realm of the I-beam's middle?

There is but one shape that can answer this call. The weak-willed square twists at its corners. The cowardly pentagon collapses under pressure. And don't get me started on that spineless turncoat known as the hexagon. Only a polygonal Superman can absorb the strain with a stoic and unyielding form.

Get me Triangle on the phone.

When you connect triangles into a single structural unit, you create a *truss* (from the French word for "bundle"). In a truss, every member experiences tension or compression. Trusses waste no material, like hunters using every part of the animal.

In ancient Egypt, the triangle labored in a vacant lot, offering a nifty trick to surveyors while the spotlight shone elsewhere. Then, millennia later and oceans away, the triangle moved from the background scenery to center stage.

5. WE BUILT THIS CITY

In the 19th and early 20th centuries, the people of North America tamed a vast continent. As this continent turned out to be rather bumpy, the project demanded bridges of every kind, from modest pedestrian footpaths to enormous railways. These bridges demanded trusses. And what do trusses demand? Triangles, of course.

The Pratt truss, devised by two brothers in 1844, consists of rows of right-angled triangles. It swept the United States, staying popular for decades.

Pratt Truss

The Warren truss, born in 1848, deployed equilateral triangles.

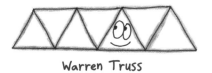

Warren Truss

The Baltimore and Pennsylvania trusses, Pratt variants with nested triangles, became common for rail bridges.

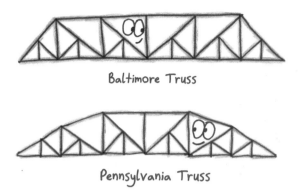

Baltimore Truss

Pennsylvania Truss

The K truss combined various triangles (while hoping no one will be reminded of the KKK).

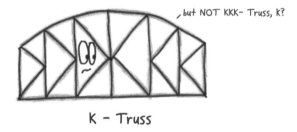

, but NOT KKK- Truss, k?

K - Truss

The Bailey truss arrived with the military needs of World War II. Its standardized modular triangles could be dismantled, shipped, and reassembled to meet the shifting urgencies of wartime.

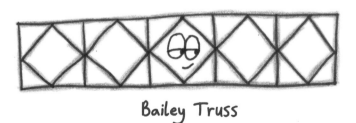

Bailey Truss

It isn't just bridges. Triangular roofs employ trusses. So do the skeletons of high-rise buildings. Heck, the standard bicycle frame is nothing but a simple two-triangle truss. To navigate a modern city is to stroll beneath triangles, to be supported by them, even to ride upon them.

Architects are hemmed in by myriad constraints: budgets, building codes, physical laws. They turn to triangles not in the spirit of artists or interior decorators, but because there are no other qualified applicants. The marriage between architecture and triangles isn't one of love. It's convenience at best, desperation at worst. So you might expect the resulting structures to strike us as last-ditch compromises, crude eyesores.

And yet, they're lovely. It is a funny paradox of design: utility breeds beauty. There is elegance in efficiency, a visual pleasure in things that just barely work.

This is, I think, the same pleasure I draw from math. A good mathematical argument, like a well-built truss, just barely works. Remove one foundational assumption, and the whole artifice collapses. There is an undeniable grace there: the minimalism, the mutually supporting elements, the utter strength without an ounce of excess.

I can't explain why I find things beautiful. (Case in point: '90s pop rock.) But I know there is something gorgeous in the narrative of the triangle. Its three-sided-ness makes it unique; its uniqueness makes it strong; and its strength makes it essential to modern architecture. Perhaps it's a stretch to claim that the triangle "saved the world," but if you ask me, it did one better. The triangle allowed the world to become what it is.

Chapter 7

IRRATIONAL PAPER

When I moved to England, I was ready to confront the shortcomings of my American upbringing. Instead of the scientific Celsius, I used the antiquated Fahrenheit. Instead of the tidy kilometer (each composed of 1000 meters), I used the idiosyncratic mile (each composed of 5280 feet). Instead of tea, I drank Starbucks-branded buckets of caramel. So I knew there'd be a rocky adjustment as I learned the customs of the civilized world.

But there was one culture shock I didn't anticipate: paper.

Like any Yank, I grew up using "letter paper." It's 8.5 inches wide and 11 inches tall, which is why you sometimes hear it called by the catchy name—really killer branding here—"eight and a half by eleven." If I'd thought about it, I'd have recognized that other countries use centimeters instead of inches, and thus letter paper was unlikely to suit them. As clunky as I find the name "eight and a half by eleven," it pales in awfulness compared to "twenty-one and three-fifths by twenty-seven and nineteen-twentieths."

Letter Paper

A4 Paper

Still, when I saw their stuff—which goes by the even less appealing name of "A4"—I developed a hot and immediate loathing. It's slightly too skinny, like a trendy pair of jeans. I was long accustomed to a loose-fit boot-cut style, and my whole being chafed at this svelte European nonsense. Letter paper has a length that's about 30% greater than its width; clearly this ratio was different. And, just as clearly, worse.

So I went to look up its measurements. I figured it'd be 22.5 centimeters by 28 centimeters, or perhaps 23 centimeters by 30 centimeters. Something nice and tidy for these metric-minded folk, right?

Nope. It's 21 centimeters by 29.7 centimeters.

What the heck?

I divided 29.7 by 21 to figure out the simplified ratio: roughly 1.41. As a math teacher, I recognized this number immediately: It's (approximately) √2, a.k.a. "the square root of 2." And in that instant, my befuddlement gave way to a snarling, smoking outrage.

√2 is **irrational**: a word that literally means "not a ratio."

The paper makers had chosen a ratio that—I can't put too fine a point on this—*is not a ratio.*

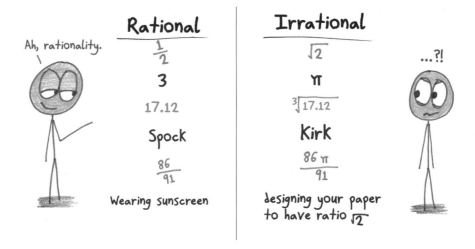

Rational	Irrational
Ah, rationality.	...?!
$\frac{1}{2}$	$\sqrt{2}$
3	π
17.12	$\sqrt[3]{17.12}$
Spock	Kirk
$\frac{86}{91}$	$\frac{86\,π}{91}$
Wearing sunscreen	designing your paper to have ratio $\sqrt{2}$

We tend to run across two kinds of numbers in life: (1) **whole numbers**, as in "I have 3 children," "My children consume 5 entire boxes of cereal each morning," and "The stains on my children's clothing range over 17 different colors"; and (2) **ratios of whole numbers**, as in "We spend 1/4 of our disposable income on Legos," "Houses with children are 17½ times more likely to be the target of Magic Marker graffiti," and "Hey, when did 2/3 of my hair go gray?"

(Your everyday decimals, I should point out, are just ratios in disguise. For example, $0.71 is simply 71/100 of a dollar.)

But some wild and exotic numbers don't fit into either category. They are not only unwhole, but unholy: They cannot be written as ratios. (My 12-year-old student Aadam, who has a far more brilliant mind than I, dubbed them the "disintegers.") No fraction or decimal that you can write down will ever quite nail these numbers. They'll always slip between the cracks.

And √2—the number that, when multiplied by itself, yields 2—turns out to be just such a number. Take a look:

Number	Number^2	So is it $\sqrt{2}$?
1.4	1.96	No
1.41	1.9881	Nope
1.414	1.999396	Nuh-uh
1.4142	1.99996164	Not quite
1.41421	1.999989924	Still no
1.414213	1.999998409	Not exactly
1.4142135	1.999999824	Yes! (Kidding. Still no.)

There's no nice decimal that quite equals √2, and there's no nice ratio, either. 7/5? Close. 141/100? Even closer. 665,857/470,832? So close you can almost taste it. But never quite right. Never quite √2.

√2 isn't merely irrational. It is, alongside π, one of the most famously irrational numbers in all of mathematics. Legend has it that the members of the ratio-worshipping cult of Pythagoras were so dismayed to find that √2 couldn't be written as a fraction that they drowned the mathematician who reported the discovery.

If European paper aspires to √2, it has chosen a goal that it can never attain. To put it in a language my British colleagues will understand: *That's bloody shortsighted, innit?*

For a few days, I lived in a state of heightened irritation. To touch this foolish paper was as loathsome as touching poison ivy or underdesk gum. I joked darkly about it, the kind of quip that's intended as charming and world-weary but contains such concentrated bitterness that people recoil to hear it.

And then, I realized I was wrong.

I didn't realize it all by myself, of course. I never do. Instead, someone pointed out to me the wonderful property that A4 paper has.

It belongs to a team.

It's exactly double the size of A5, quadruple the size of A6, and eight times the size of the adorably tiny A7. Meanwhile, it's exactly half the size of A3, a quarter the size of A2, and one-eighth the size of the impressively ginormous A1.

A1	A2 A2	A3 A3 / A3 A3	A4 × 8	A5 × 16

As befits the individualistic culture that employs it, letter paper is an island. Our 8½" × 11" size bears no particular relationship to the smaller and larger paper types in circulation. It's a one-off.

How... American.

By contrast, global paper is as interconnected as the globe itself. A4 belongs to a unified series of papers: all different sizes but all exactly the same proportion.

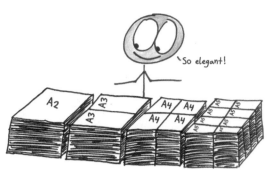

So elegant!

If you are a serious user of paper—like, say, a math teacher—this is delightful. You can shrink down two pages to fit perfectly on a single sheet. Or you can put two pages together to fill a sheet of A3. For those unfeeling and indifferent people who get their fun at beaches, nightclubs, and French restaurants, this is perhaps not very stimulating. But for a stationery aficionado, this is a transcendent thrill. *This paper makes sense.*

And once I understood this, I saw that the blunder was no blunder at all. It was inevitable: the only way to make this magical Russian nesting doll system of paper work properly.

To see why, let us imagine the scene.

INT. PAPER LABORATORY - NIGHT

Gwen and Sven, two beautiful paper scientists, are working on a top-secret research project. It is code-named "The Prince and the Paper," or perhaps "Paper Caper"—whichever sounds cooler in their vague, foreign accents. It is late. They are exhausted but devoted to their work.

GWEN: All right, Sven. I may not know our precise nationalities, but I know one thing: The fate of civilization depends on our ability to create a series of papers where folding one size in half gives you the next size in the series.

SVEN: The stakes could not be higher. But . . . what dimensions will such paper have?

GWEN: There's only one way to find out.

With a decisive motion, Gwen folds a piece of paper in half, and marks three lengths: "long" (the length of the original paper), "medium" (the width of the original paper) and "short" (the width of the half-sized paper).

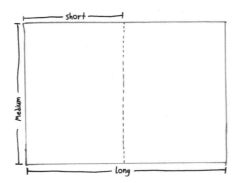

GWEN (continuing): Now, what's the ratio between "long" and "medium"?

SVEN: That's what we're trying to find out.

GWEN: Well, what's the ratio between "medium" and "short"?

SVEN: Blast it, Gwen! We know that's the same ratio, but we still don't know what it is.

A moment passes, filled with romantic tension.

GWEN: Okay. Let's say that the "medium" is *r* times longer than the "short."

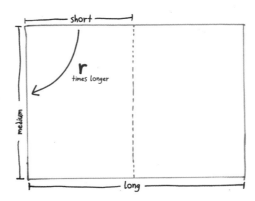

SVEN: But what's *r*?

GWEN: I don't know yet. I only know that it's more than one but less than two, because the "medium" is longer than the "short," but it's not twice as long.

SVEN: All right. And I suppose that makes the "long" also *r* times longer than the "medium."

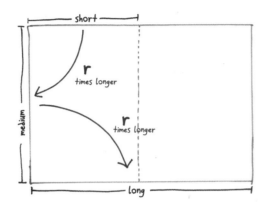

GWEN: So if you want to turn "short" into "long," you'd have to multiply by *r* (to get "medium") and then multiply by *r* again. That's *r squared.*

SVEN (slapping the table): You genius with perfect pores and inexplicable high heels—Gwen, you've done it!

GWEN: Have I?

SVEN: The "long" is r^2 times longer than the "short." But look: it's also *double* the length of the short!

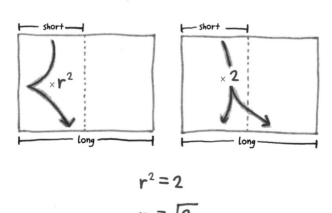

$$r^2 = 2$$
$$r = \sqrt{2}$$

GWEN: My word . . . you're right . . . which means . . .

SVEN: Yes. r^2 is 2.

GWEN: So *r* is the square root of 2! That's the secret ratio that will end suffering and unify humanity!

SVEN (suddenly changing accents): All right, Gwen. Hand over the ratio.

GWEN: Sven? Why are you holding that gun?

Contrary to initial appearances, the makers of A4 paper didn't choose this ratio to upset me personally. Nor did they choose it in deference to arbitrary fashions, or in obstinate defiance of American hegemony, or even for the sadistic pleasure of selecting an irrational ratio.

In fact, they didn't really choose it at all.

What they chose was to create a system of paper where folding one size in half yields the next size in the series. That's a pretty cool and hard-to-fault feature. But once they committed to this path, the decision was out of their hands. There is only one ratio that achieves this property, and it just so happens to be the famously irrational √2.

Now, I know we all like to imagine paper designers as unshackled fantasists, bound only by the limits of their imagination. The reality is actually more interesting. Designers move through a space of possibilities that are governed by logic and geometry. This landscape has immovable features: some numbers are rational, some aren't, and there's nothing a designer can do about it. Instead, the designer must navigate around these obstacles—or, better yet, turn them into assets, like an architect whose building is in harmony with the surrounding environment.

Long rant made short: I've come around on A4 paper. Now that I know *why* they're aiming for √2, the fact that they're doomed to miss the target microscopically doesn't much bother me. To be honest, A4 paper doesn't even look wrong to me anymore. Now it's letter paper that looks off: a little stout and old-fashioned.

It seems that I've completed the transition from one type of insufferable American to another. From jingoistic defender of my own arbitrary national customs, I've become a fervent evangelist for arbitrary foreign customs. I'm even drinking fewer caramel buckets these days, although— much like letter paper—I'm sure I'll never give it up altogether.

Chapter 8

THE SQUARE-CUBE FABLES

JUST-SO STORIES OF MATHEMATICAL SCALING

Fables and math have a lot in common. Both come from dusty, moth-eaten books. Both are inflicted upon children. And both seek to explain the world through radical acts of simplification.

If you want to reckon with the full idiosyncrasy and complexity of life, look elsewhere. Ask a biologist, or a painter of photorealistic landscapes, or someone who files their own taxes. Fable tellers and math makers are more like cartoonists. By exaggerating a few features and neglecting all the rest, they help explain why our world is the way it is.

This chapter is a brief collection of mathematical fables. They show how diverse settings, from baking to biology to the financing of the arts, are governed by the constraints of geometry. At these tales' collective heart is a single fundamental idea, a moral so simple that even Aesop neglected to spell it out: size matters.

A big statue is not just a big version of a small statue. It is a different object entirely.

1. WHY BIG PANS MAKE BETTER BROWNIES

You and I are baking. We stir the batter with pride, bettering mankind via chocolate miracle. But when the oven is preheated, the cupboards defy us. The only available pan has double the dimensions of the one suggested by the cookbook.

What do we do?

Needing to fill a double-sized pan, we are tempted to double the recipe. But this would be a mere half measure. Look carefully and you will see: We need to *quadruple* the recipe.

What happened? Well, the flat pan has two dimensions: length and width. By doubling the length, we doubled the pan's area. And by doubling its width, we doubled the area *again*. That means the area has been double-doubled. Hence, a multiplier of four.

This happens any time that you grow a rectangle. Triple the sides? Nine times the area. Quintuple the sides? Twenty-five times the area. Multiply the sides by 9 bajillion? That'll yield 81 bajillion bajillion times the area.

Or, more precisely: **Multiply the lengths by r, and you multiply the area by r^2.**

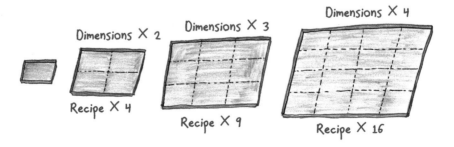

It's not just rectangles. The same principle holds for all two-dimensional shapes: trapezoids, triangles, circles, and any other vessel into which you might pour the sacred brownie batter. As the length grows, the area grows much faster.

Back in the kitchen, we have just finished quadrupling our recipe when a forgotten cupboard reveals the pans we'd been looking for all along. We blame each other, then laugh, because who can stay mad when chocolate glory is so close at hand?

We now face a choice: Shall we bake the brownies in the one big pan, or in four small ones?

This is a fable, so we shall ignore the details. Forget oven temperature, cooking times, heat flow, and minimizing the dishes to clean. Focus instead on one matter: size itself.

As a brownie pan grows, its exterior (the length of the edges, which is one-dimensional) gets larger. But its interior (which is two-dimensional) gets larger faster. This means that small shapes tend to be "edge-heavy" while large shapes are "interior-heavy." In our case, the four small pans will have the same area as the single big one, but they'll have *double* the edge length.

The small pans will maximize our quota of edge brownies, while the big pan will minimize it.

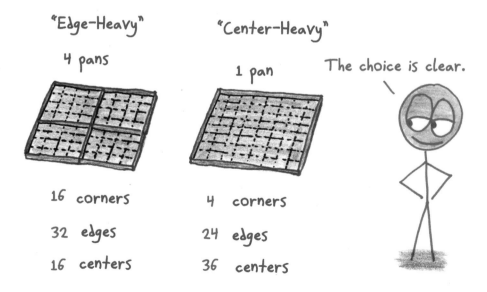

"Edge-Heavy"

4 pans

16 corners

32 edges

16 centers

"Center-Heavy"

1 pan

4 corners

24 edges

36 centers

The choice is clear.

No matter how I try, I cannot fully embrace the humanity of people who prefer edge brownies. Who would forsake a fudgy wonder for a chewy chore, a crispy mistake? I can only imagine that they would prefer bones over meat, crumbs over crackers, side effects over pain relief. Such people lie beyond explanation and exculpation. Go big pan, or go home.

2. WHY THE AMBITIOUS SCULPTOR WENT BANKRUPT

About 2300 years ago, the Greek island of Rhodes repelled an attack from Alexander the Great. In a mood of self-congratulation, the people commissioned local sculptor Chares to build a magnificent commemorative statue. Legend tells that Chares originally planned a 50-foot bronze sculpture. "What if we supersize it?" said the Rhodians. "You know, double the height? How much would that cost?"

"Double the price, of course," said Chares.

"We'll take it!" said the Rhodians.

But as construction unfolded, Chares found his funds drying up. The material expenses astounded him; they far exceeded his budget. Bankruptcy loomed. According to some, Chares chose to escape financial ruin by taking his own life, and never saw the masterpiece finished. But perhaps, before he died, he understood his mistake.

He was too shy about jacking up the price.

To see why, forget all the details. Pay no mind to the labor market for Greek construction workers, or the pricing of wholesale bronze. Heck, forget artistry altogether: imagine Chares was just building a giant bronze cube. We must focus on a single overpowering question: size.

What happens when you double the dimensions of a 3D shape?

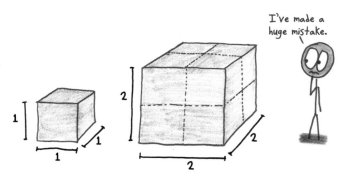

Well, you've doubled the length; that doubles the volume. You've doubled the height, which doubles the volume again. And you've doubled the depth, which doubles the volume a third time. That's a triple double, although not in the Russell Westbrook sense: here, double-double-double amounts to "multiply by eight."

The results are as clear as they are startling: Volumes get big *fast*. Triple the sides of a cube? That's 27 times the volume. Multiply the sides by 10? You'll multiply the volume by an absurd factor of a *thousand*. And what's true of cubes is true of every shape: pyramids, spheres, prisms, and (unlucky for Chares) exquisite statues of the sun-god Helios. In precise terms: **Multiply the length by r, and you multiply the volume by r^3.**

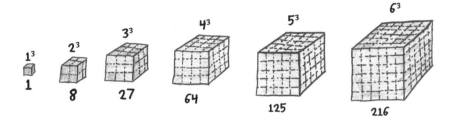

If Chares had created a one-dimensional artwork (*The Colossally Long String of Rhodes*), then the aesthetics might have suffered but his price scheme would have worked fine: doubling the length would double the required material. Or suppose he'd been commissioned to create a two-dimensional painting (*The Colossal Portrait of Rhodes*). He'd still be lowballing the contract, but less so: doubling the height of a canvas quadruples its area, demanding four times the paint. Alas, Chares had the misfortune to work in a full three dimensions. Doubling the height of his statue multiplied the bronze requirements by eight.

When a 1D length grows, a 2D surface grows faster, and a 3D volume grows faster still. The Colossus of Rhodes—one of the wonders of the ancient world—doomed its own creator for the simple reason that it was three-dimensional.

double materials

quadruple materials

octuple materials

10

20

30

3. WHY THERE AREN'T GIANTS

King Kong, the three-story ape. Paul Bunyan, the lumberjack whose footsteps carved out lakes. Shaquille O'Neal, the mythical seven-foot one, 325-pound basketball player who could do everything except make free throws. You know these stories, and you know just as well that they are fantasies, legends, wide-eyed fictions. There's no such thing as giants.

Why? Because size matters.

Suppose we take exemplary human specimen Dwayne Johnson and double his dimensions. Now with double the height, double the width, and double the depth, Dwayne's total body mass has grown by a factor of eight.

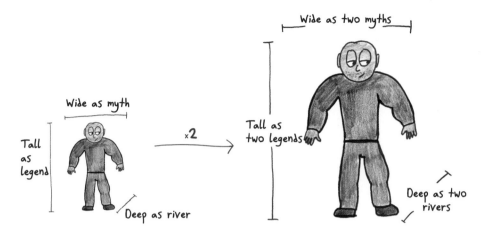

So far, so good. But take a glance at his legs. To keep our man standing, his bones will need eight times the strength. Can they muster it?

I doubt it. They've undergone two helpful doublings (in width and depth) but one useless doubling: length. Just as you don't reinforce a column by making it taller, you can't bolster a leg by making it longer. Extra length confers no extra strength, just extra strain, as the bone's base must now support the greater mass above.

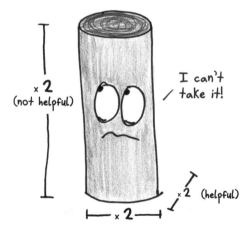

Dwayne's leg bones won't keep pace with the demands placed on them: A factor of four can't match a factor of eight. If we keep growing Dwayne Johnson, doubling and tripling and quadrupling his size, then eventually he will reach a critical breaking point. His limbs will buckle and splinter beneath the overwhelming weight of his torso.

We're dealing with a process called *isometric scaling*: growing a shape while preserving its proportions. (*Iso-* means "same"; *-metric* means "measurement.") It's a lousy method for making big animals. Instead, we need *allometric scaling*: growing a shape while altering its proportions accordingly.

When an animal's height grows by 50%, its legs can keep pace with the strain only by growing 83% thicker. That's why cats can survive on slender limbs, whereas elephants need pillars to stand on.

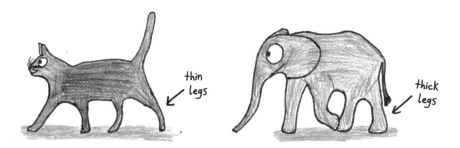

A constraint upon Dwayne Johnson is a constraint upon us all, and thus, giants shall always belong to the realm of myth. Paul Bunyan's tibiae would shatter with every lake-punching step. King Kong's muscles (their strength growing with the square of his height) could never carry his bulk, which grows "with the cube": he'd merely sit there, a giant monkey puddle suffering perpetual heart failure. And Shaquille O'Neal? Well, his tale is so implausible that I doubt anyone truly believes it.

4. WHY NO ANT FEARS HEIGHTS

My Nightmare
(actual size)

Ants are horrifying. They lift objects 50 times their body mass, work together in flawless coordination, and thrive on every corner of the planet. This global army of weightlifting pincer-jawed telepaths outnumbers humans by a factor of millions. Visions of their alien faces would keep me up at night if not for one redeeming fact:

Ants are very, very small.

It's time now to solidify the pattern of our earlier fables. When a shape's length grows, its surface area grows faster, and its volume grows faster still.

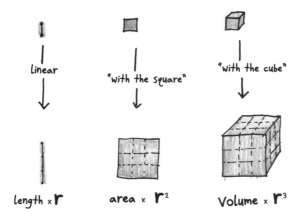

linear   "with the square"   "with the cube"

length x r area x r^2 Volume x r^3

This means that big shapes (like humans) are "interior-heavy." We have lots of inner volume per unit of surface area. Small shapes (like ants) are the opposite: "surface-heavy." Our formic nemeses have lots of surface area relative to their tiny interior volume.

What is it like being surface-heavy? First, it means you never need to fear heights.

When you fall from a great height, two forces play tug-of-war: downward gravity and upward air resistance. Gravity works by pulling on your mass, so its strength depends on your interior. Air resistance works by pushing against your skin, so its strength depends on your surface.

In short, your mass speeds up your fall, while your surface area slows it down. That's why bricks plummet and paper flutters, why penguins can't fly and eagles can.

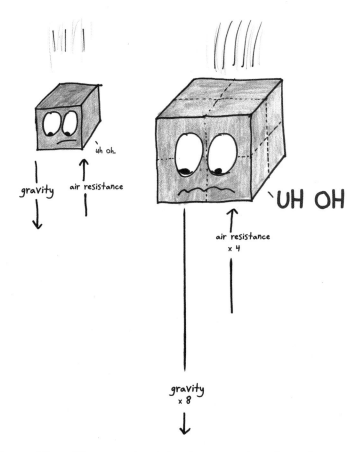

You and I are like penguins: lots of mass, not much surface area. When falling, we accelerate to a top speed of nearly 120 miles per hour, which is a rather disagreeable way to hit the ground.

By contrast, ants are like paper eagles: lots of surface area, not much mass. Their terminal velocity is just 4 miles per hour. In theory, an ant could leap from the top of the Empire State Building, land with all of its six feet on the pavement, and walk away singing "Ants Marching."

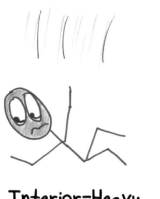

HE wakes up in the morning...

| Interior-Heavy | Surface-Heavy |

So is being surface-heavy all fun, games, and parachute-free skydiving? Of course not. Ants have woes, too, and by "woes" I mean "a debilitating and wholly justified fear of water."

The trick is surface tension. Water molecules love sticking together, and they're willing to defy small amounts of gravity to do it. Thus, when any object emerges from a bath, it carries with it a thin layer of water, about half a millimeter thick, held in place by surface tension. For us, that's a trifle: a pound or so, less than 1% of our body weight. We towel off and go on living.

Compare that to the ordeal of a bathing mouse. The water layer's thickness remains the same, about half a millimeter, but our rodent friend finds this a far greater burden. The surface-heavy mouse steps out of the tub carrying bathwater that equals its own body weight.

And for an ant, the situation is dire. The clinging water outweighs its body by an order of magnitude; a single wetting can be fatal. That's why ants fear water the way I fear them.

5. WHY BABIES NEED BLANKIES

Although the cutting edge of parenting advice evolves minute by minute, some principles never change: cuddles are good; head trauma is bad; and swaddling your little one is a must. We've been wrapping our young since the Paleolithic, and thousands of years from now, I'm sure that the survivors of the zombie apocalypse will still be swaddling their traumatized infants.

Babies need blankets because—please forgive the technical jargon—babies are small.

Once again, ignore the details: the tiny toothless mouth, the eensy wiggling toes, the small bald head that smells so amazing. Think of a baby the way you'd think of any organism: as a homogenous bundle of chemical reactions. Every activity of the body is built on such reactions; in some sense, the reactions *are* the creature. That's why animals are so temperature-sensitive: Too cold, and the reactions slow to a halt; too hot, and some chemicals deform, making key reactions impossible. You've got to keep a close eye on the thermostat.

Heat is created by reactions in each cell (i.e., in the interior). And heat is lost through the skin (i.e., at the surface). This creates a familiar tug-of-war: interior vs. surface.

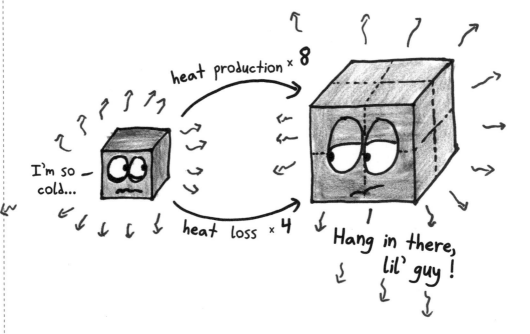

Bigger animals, being more interior-heavy, will have an easy time keeping warm. Smaller ones, being surface-heavy, will struggle. That's why you're most vulnerable to cold in your surface-heavy extremities: fingers, toes, and ears. This also explains why cold climates support only big mammals: polar bears, seals, yaks, moose, mooses, meeses, and (depending on your zoology professor) the Sasquatch. A surface-heavy mouse wouldn't stand a chance in the Arctic. Even at moderate latitudes, mice cope with heat loss by eating a quarter of their body weight in food each day.

A baby isn't a mouse, but it's definitely not a yak. Its tiny body expends heat like governments spend money. And to stifle that heat loss, there's no cuddlier option than a blankie.

6. WHY AN INFINITE UNIVERSE IS A BAD IDEA

Once you start looking for them, square-cube fables are everywhere. Geometry governs every design process, and no one can defy its rules—not sculptors, not chefs, not the forces of natural selection.

Not even the cosmos itself.

Perhaps my favorite square-cube fable of all is the "paradox of the dark night sky." It dates to the 16th century. Copernicus had just laid out the notion that Earth was not the center of existence, that ours is an ordinary planet orbiting an ordinary star. Thomas Digges, importing Copernicus's work to England, took a further step. He proposed that the universe must go on forever, a sparse and ageless cloud of far-flung stars, extending into infinity.

Then Digges realized that if that were true, the night sky should be a blinding, scorching white.

To see why, we're going to need a magical pair of sunglasses. I call 'em "Digges shades." These sunglasses have an adjustable dial and a wondrous property: they block out all the light from beyond a certain distance. For example, set them to "10 feet," and most of the world goes dark. Sunlight, moonlight, streetlamps—all of these vanish, leaving only the light sources within 10 feet of your current position. A reading lamp; an iPhone screen; perhaps nothing more.

Let's tune our Digges shades to "100 light-years." Gazing up at the night sky, we say goodbye to Rigel, Betelgeuse, Polaris (a.k.a. the North Star), and many other familiar orbs. The remaining stars, those within a hundred-light-year radius, number about 14,000—a sparser, dimmer heavens than the usual.

Altogether, these stars will have some total brightness. We'll give a lowball estimate according to the following equation:

$$\text{Total Brightness} = \text{Number of Stars} \times \text{Minimum Brightness}$$

Turn the dial on your Digges shades, doubling the distance: 200 light-years. Stars reappear. The sky brightens. But by how much?

Well, the visible sky forms a 3D hemisphere around Earth. By doubling its radius, we've octupled its volume. Assuming (as Digges did) that the stars follow an even distribution, like trees in a forest, then we're now seeing eight times as many stars as before. Their population leaps from 14,000 to more than 100,000.

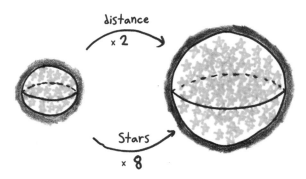

distance
× 2

Stars
× 8

But these new stars are farther, which means they'll be dimmer. The question, once again, is: By how much?

Each star appears in the night sky as a tiny circle. The bigger the circle, the more light hits our eye. Since this is a fable, we can ignore personality differences among stars—their temperatures, their colors, whether they have cool names like Betelgeuse and Polaris. We'll assume, like unapologetic star-racists, that all stars are identical. The only thing that matters is their distance.

If Star A is twice the distance of Star B, then its circle will be half the height and half the width. That leaves it with a quarter of the area—and thus, a quarter of the brightness.

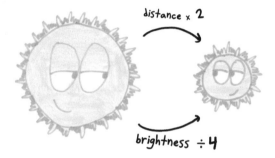

So, what can we make of our new and enlarged night sky? The number of visible stars has been multiplied by eight; their minimum brightness has been divided by four; and (correct me if I'm wrong) eight divided by four is two, meaning that the total brightness, according to our simple equation, has doubled.

Double the radius on your Digges shades? Double the brightness of the sky.

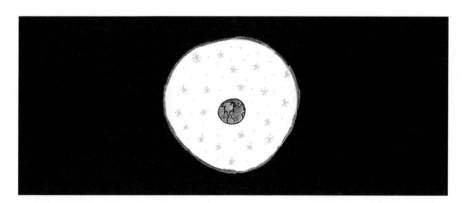

Because "number of stars" is 3D and "minimum brightness" is 2D, the sky becomes brighter and brighter the more we spin our dial. Triple the radius, to 300 light-years? You'll triple the original brightness. Multiply the radius by a thousand? Our night sky is now a thousand times brighter than before.

You can see where this is going—or rather, you can't, because before long you are literally blinded by the stars. The night sky can grow a million, a billion, a trillion, even a googol times brighter than the initial level. Given a large enough radius, the stars outshine the day and overpower the sun, turning the sky into a scorching miasma of infinite temperature. Compared to this nonstop vaporizing blast of unfathomable brightness, the dinosaurs got off easy with a measly asteroid.

Take off your Digges shades, and the totality of the night sky will brighten you to death.

If the universe is infinitely large, why isn't our night sky infinitely bright? This paradox endured for centuries. In the 1600s, it nagged at Johannes Kepler. In the 1700s, it spurred Sir Edmund Halley. In the 1800s, it inspired an Edgar Allan Poe prose poem that the author called his greatest work. (Critics disagreed.)

Only in the 1900s did the paradox find a happy resolution. The deciding factor is not size, but age. Whether or not the universe is infinitely big, we know for certain that it is finitely old, born less than 14 billion years ago. Thus, any settings on your Digges shades beyond "14 billion light-years" are useless. The light hasn't had time to reach us, so there is nothing at that distance to block out.

This square-cube fable is more than just a clever argument. It was an early line of evidence for the big bang theory. To me, this is the apotheosis of the square-cube spirit. By thinking about the simplest aspects of our universe—what's 2D, what's 3D—we can reach sweeping levels of understanding. Sometimes it takes an act of radical simplification to see the world as it truly is.

Chapter 9

THE GAME OF DICE

FOR 1 TO 7,500,000,000 PLAYERS

Thank you for purchasing the Game of Dice! This fun pastime for the whole civilization is suitable for ages from "stone" to "digital," and beloved by gamers from the commoner to the tyrant. Don't take my word for it! Just ask these Roman emperors:

I wrote a book called "how to win at dice."

Claudius

My palace featured special dicing rooms!

Commodus

I gave every dinner guest a bit of money, in case they wanted to gamble on dice during the evening!

Augustus

Dice? Oh yeah, I cheated constantly.

Caligula

This instruction manual will introduce you to the basic rules. The Game of Dice is equal parts theory and practice, stretching the mind and the fingertips alike. Let's play!

THE GOAL OF THE GAME:
Enthrall and entertain humankind with a device that generates outcomes beyond its control.

We begin with a class of characters known as "people." These creatures love being in control, which is why they invented cars, guns, governments, and central air-conditioning. But they also obsess over what's out of their control: traffic, weather, their children, and the success of famous guys who play sports for money.

In their hearts, they wish to confront fortune, to hold their own powerlessness in their palms. This is where dice come in. They are handheld pieces of fate.

In 6000 BCE, ancient Mesopotamians tossed stones and seashells. Ancient Greeks and Romans rolled the knucklebones of sheep. The first Americans threw beaver teeth, walnut shells, crows' claws, and the pits of plums. In the Sanskrit epics of ancient India, kings cast handfuls of hard bahera nuts. These natural dice enabled games of chance, the telling of fortunes, the dividing of prizes, and (no doubt) other customs ranging from the sacred to the mundane. Like desserts or naps, the idea of dice was so obvious and beautiful that every society came to it on its own.

Nowadays, only a few die-hard traditionalists play Monopoly with beaver teeth. Civilizations have advanced from "found" dice to designed ones. This is where the true Game of Dice begins.

RULE #1: GOOD DICE PLAY FAIR.

When you roll a die, each face should have an equal chance of coming up. Otherwise, people get cranky and suspicious and stop inviting you to their backgammon parties.

A useful starting point: *congruence.* Two shapes are congruent if one can be superimposed upon the other. Congruent shapes are identical twins, in that their angles and edges all match. Thus, your first idea for a fair die might go something like: *make sure the faces are congruent.*

This sounds great . . . until you meet the **snub disphenoid**.

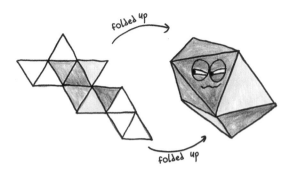

This snout-faced mole rat of a polyhedron dashes our hopes. Its 12 faces are equilateral triangles, all identical. Yet it does not make a fair die. At some corners, four triangles meet. At others, five do. When you roll the little monster, some faces come up more than others. I'm afraid congruence just isn't enough.

We need *symmetry.*

In ordinary speech, "symmetry" refers to a vague, pleasing uniformity. Its mathematical meaning is more specific: a geometric action that transforms an object without really changing it. For example, a square table has eight symmetries:

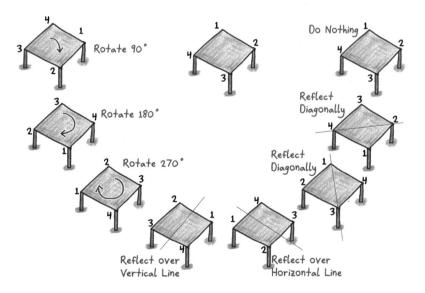

The symmetries leave the table unchanged, though on careful inspection, they scramble the locations of the corners. For example, the 180° rotation swaps pairs of opposite corners: #1 and #3 trade places, as do #2 and #4. Contrast this with the diagonal reflection, which swaps #2 and #4, but leaves #1 and #3 where they are. Symmetries on dice work in much the same way: they rearrange the faces while leaving the overall shape unchanged.

Symmetry offers a surefire path to fair dice. Just pick a shape with enough symmetries that every face can be swapped with every other.

One example: the **dipyramid**. Take two identical pyramids and glue their bases together. The right combination of rotations can swap any triangular face with any other, meaning the faces are geometric equivalents, and so the die is fair.

Another example: the **trapezohedron**, which looks like a dipyramid with some nifty carving at its equator, to turn the three-sided triangles into four-sided kites.

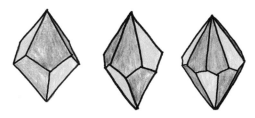

With a dipyramid or a trapezohedron, you can create any even-sided die: eight-sided, 14-sided, 26-sided, 398-sided. In theory, every single one will be fair, each face appearing with equal likelihood. You might think that we've solved the problem. The Game of Dice is over, yes?

Not so fast! People are fickler than that. It's not enough for dice to *be* fair . . .

RULE #2: GOOD DICE *LOOK* FAIR.

We've met dice candidates that (1) are easy to define, (2) produce fair outcomes, and (3) come adorned with awesome Greek and Latinate names. Yet these promising models prove historical duds, the failed presidential nominees of the randomization world. To my knowledge, no dice-rolling culture has embraced the dipyramid, and only one welcomes the trapezohedron: the shadow society known as "Dungeons & Dragons players," whose 10-sided die (or d10) is trapezohedral.

Why so finicky, humankind? How can you spurn good shapes, cast fair dice aside?

Roll a skinny dipyramid, and you'll see the problem. It doesn't tumble. Instead, stabilized by its two pointy ends, it just sort of rolls, like a crooked paper-towel tube, rocking a little back and forth between the two clusters of faces. Its stops are precarious; a jokester's breath could tip it from one face to another. That's a recipe not for a fun game of Parcheesi, but for a family-shredding afternoon of recriminations.

The best dice need more than symmetric faces. They need symmetric *everything*. And if you're a polyhedron aficionado, you know what that means.

Platonic solids!

Of all the straight-edged 3D shapes, the Platonic solids are the most perfect. Their symmetries can swap any two faces, any two corners, or any two edges—symmetries so magnificent that even a hardened cynic cannot doubt their eminent fairness.

There are five Platonic solids: no more, no less. And each member of this geometric pantheon has graced Earth in the form of a die.

1. The **tetrahedron**, a pyramid of equilateral triangles. In 3000 BCE, Mesopotamians rolled tetrahedra to play the Royal Game of Ur, an early precursor to backgammon.

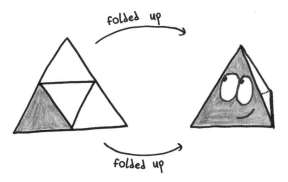

2. The **cube**, a prism of squares. Simple, sturdy, and easy to manufacture, it remains history's most popular die. Our oldest specimen, a cube of baked clay excavated in northern Iraq, dates to 2750 BCE.

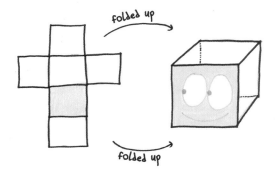

3. The **octahedron**, a special dipyramid of equilateral triangles. Egyptians sometimes entombed such dice alongside their dead.

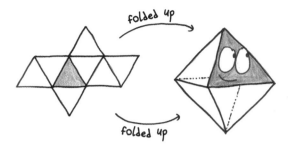

4. The **dodecahedron**, a pleasing gemstone of 12 pentagonal faces. They helped tell fortunes in 1500s France. Astrologers today enjoy their correspondence to the 12 signs of the Zodiac.

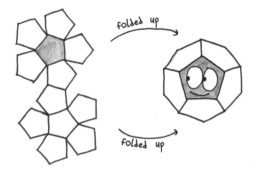

5. The **icosahedron**, a figure made of 20 equilateral triangles. Though ubiquitous in Dungeons & Dragons, it enjoys greatest favor among fortune-tellers. That includes Magic 8 Balls, which are icosahedra floating in water. If you've ever shaken one, then you've asked a Platonic solid about your future.

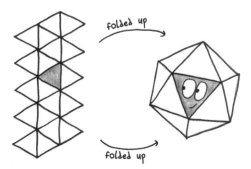

In the Game of Dice, the Platonic solids are the trump cards. One cannot picture them without heavenly choirs piping up in the background. But, as an exclusive pantheon of five, they are all too rare. They let us randomize among 4, 6, 8, 12, or 20 outcomes . . . but no others.

It's worth trying to break out of the box. Why not a paradigm-busting design, a fresh and innovative way to randomize any number of outcomes?

Spoiler: It's harder than it sounds.

RULE #3: GOOD DICE
WORK WELL ANYWHERE.

One alternative is "long dice." Instead of worrying about every face having an equal chance, build a long prism instead.

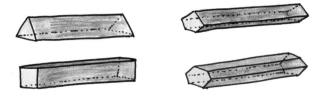

These dice work not because all faces come up with equal likelihood, but because two of them never come up at all. Long dice play fair, look fair, and allow you to randomize among any number of outcomes. So why aren't they more popular?

Well . . . they roll too much.

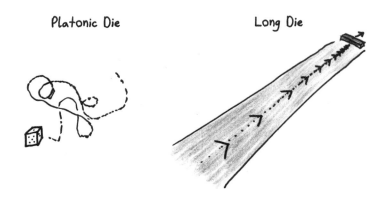

Platonic Die   Long Die

Whereas Platonic solids treat the table like a dance floor, leaping this way and that, long dice roll in a single direction. You've got to clear a whole bowling-alley path for them. What kind of self-important die demands a red carpet?

Back to the chalkboard we go, to recruit another mathematical principle: *continuity*.

Grab your long die from wherever it rolled to a stop. I'll wait. (And wait. And wait.) Now, as we can see, the two bases almost never come up. But imagine shortening the die, to make a "less long" variant. The shorter it grows, the more probable the two bases become. After a while, you're left with a die so short that it's basically a coin, at which point the roles reverse. Now, almost every toss lands on a base.

Somewhere in the middle, there is a moment of crossing over: a length at which the bases and the lateral faces enjoy the same likelihood. Somewhere in the middle is a fair die.

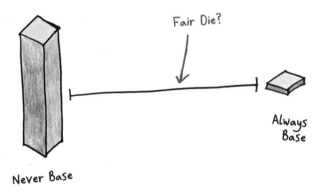

In theory, you can pull this trick with any polyhedron, achieving shapes that are funky-looking yet fair. So where are they? Why aren't novelty shops selling nifty trick dice that look wild but play well—the bizarro antithesis of loaded dice?

It's because the adjustments are too sensitive. Fair on hardwood? Unfair on granite. Fair at one size? Unfair when built double-sized. Fair on one roll? Unfair at a different strength, or with a different spin. Change any condition, no matter how incidental, and you alter the physics. Such dice would be chained to the hyperspecific circumstances of their birth. People want portable, durable dice—not fussy prima donnas.

RULE #4: THEY'RE SIMPLE TO ROLL.

Say we want a random letter from the 26 in the alphabet. The icosahedron has too few faces. The dipyramid wobbles funny. The long die rolls into the distance like a runaway meatball. Ruling out those options, we're stuck. Is there no way to accomplish the simple task of picking a random letter?

Sure there is. Just flip five coins.

There are 32 outcomes, all equally likely. Assign a letter to each of the first 26, and interpret the leftovers as "Start again."

A HHHHH	**I** HTHHH	**Q** THHHH	**Y** TTHHH
B HHHHT	**J** HTHHT	**R** THHHT	**Z** TTHHT
C HHHTH	**K** HTHTH	**S** THHTH	
D HHHTT	**L** HTHTT	**T** THHTT	TTHTH
E HHTHH	**M** HTTHH	**U** THTHH	TTHTT
F HHTHT	**N** HTTHT	**V** THTHT	ReFlip TTTHH
G HHTTH	**O** HTTTH	**W** THTTH	TTTHT
H HHTTT	**P** HTTTT	**X** THTTT	TTTTH
			TTTTT

This procedure works for any randomization scenario. Say we want to pick a random word from the Lord of the Rings trilogy. There are about 450,000 to choose from. So flip 19 coins, yielding more than 500,000 possible outcomes. Assign a word to each. If you land on one of the unassigned outcomes, just flip all 19 coins again.

Heck, you don't even need 19. Just flip a single coin 19 times.

When HHHHHHHHHHHHHHHHHHH

Mr. HHHHHHHHHHHHHHHHHHT

Bilbo HHHHHHHHHHHHHHHHHTH

Baggins HHHHHHHHHHHHHHHHHTT

of HHHHHHHHHHHHHHHHTHH

. . .

This is even more tedious than actually reading Lord of the Rings.

By this logic, there is nothing dice can accomplish that a single coin cannot match. And yet it's hard to picture a Vegas casino drawing in crowds with a coin-only craps table or a penny-based roulette wheel.

The problem is obvious: these systems are too complicated. It's an awful lot of trouble to record a sequence of coin flips, look up the result in an index, and then (if necessary) repeat the whole process. You want a single roll. No sides wasted. No user manual required.

This principle nixes some clean mathematics. For example, you could randomize among four outcomes with a simple cubical die. Just label two of the faces "roll again." But this approach bugs people. It feels inelegant to waste faces. When sharing a cake among four friends, you would never cut six pieces and then throw out the two extras.

I don't like it. What if it says "roll again" like 10 times in a row?

yes, but WHAT IF

The probability of that is 1 in 59,000.

I suspect that's why Dungeons & Dragons players roll a four-sided die instead. I read this as a sign of desperation, because of all the Platonic solids, the tetrahedron is history's least popular die. Again, any human can see why: It lands with a face down but not a face up. This feels wrong, like asking someone to guess the number that you're *not* thinking of.

Across the millennia, humans have shunned the tetrahedron in favor of dice that have parallel pairs of opposite faces, so that every "down" face corresponds to an "up." Mathematics doesn't care. But humans are the ones calling the shots.

RULE #5: THEY'RE HARD TO CONTROL.

Remember the whole purpose of the die? It puts the human body in contact with higher forces: random chance, karmic fate, the will of the gods. This enables brainstorming, games of luck, the telling of fortunes, and other profound expressions of our humanity.

So of course, people try to cheat.

One path: manipulate the exterior of the die. For example, extend it by an undetectable increment, forming a brick shape. Or create faces with slight bulges (making them less likely) or indentations (making them likelier). Or you can cap some faces with a bouncy material, or sandpaper them down so the die prefers them. These tricks are as old as ruins. I mean that literally: loaded dice with shaved corners have been excavated in Pompeii.

A second approach: manipulate the *interior* of the die. "Trappers" are dice with two hidden compartments; the right motion will move a heavy blob of liquid mercury from one compartment to the other, shifting the probabilities. (If poison metals aren't your cup of tea, then use wax that melts just below body temperature.) Another scheme: Back when wooden dice were popular, some cheaters grew small trees with pebbles embedded in the branches. Then they'd carve a die from the wood, leaving that invisible rocky weight within. This con job demands not just extraordinary patience but above-average botanical skill.

A third approach: renumber the faces. A normal die features opposite faces that sum to seven. (The pairs are 1 and 6, 2 and 5, and 3 and 4.) In cheat dice known as "taps," some numbers are duplicated, yielding (for example) opposite pairs of 6 and 6, 5 and 5, and 4 and 4. From any angle, your victims will see only three faces, so nothing appears to be wrong.

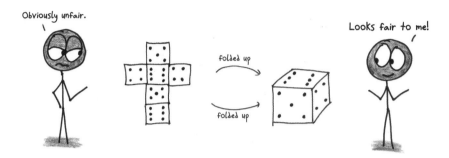

Though all of these cheating methods target cubical dice, it's not that cubes are extra vulnerable. They're just extra popular. Evidently, there's more money in craps than in D&D.

RULE #6: GOOD DICE FEEL NICE.

The Game of Dice is like most games, in that nobody really needs it. We live in the 21st century. I jetpack to work and take vacations by flying car. Okay, I don't, but I've got half the world in my pocket in the form of a 5-ounce computer. Technology is making us all obsolete, and that includes the blue-collar workers called dice. Watch: I'm going to stop writing and go simulate a million rolls of a cubical die in Microsoft Excel. I'll let you know how long it takes.

Okay, done. Took about 75 seconds. Here are the results:

Number	Frequency
1's	166,335
2's	166,598
3's	167,076
4's	166,761
5's	167,103
6's	166,127

Not only is computer randomization faster and easier than throwing a plastic cube across a table, it's more random, too. Casinos could eliminate craps tables and roulette wheels tomorrow, and their digital replacements would outperform the old dinosaur randomizers by leaps and bounds.

But what fun would that be?

Dice are meant to be held. The first time I played Dungeons & Dragons (okay, the only time), I encountered something more alluring than any of the orcs and mages in the game itself: the Zocchihedron, the 100-sided die. Can you imagine? A hundred sides! A die that rolls for 30 seconds before coming to a stop! I knew that two d10s (one for the tens digit, one for the units) served better and was fairer than Lou Zocchi's lumpy shaved golf ball. I didn't care. I wanted to roll that d100.

Greeks must have felt that same allure when tossing the sheep knuckle they called the *astragalos*. They valued its four sides at strange values (1, 3, 4, and 6) and threw them by the handful. A throw of all 1's was "the dogs," the worst throw imaginable. The best throw (either all 6's or one of each face, depending who you ask) was called "Aphrodite." The astragali weren't fair; they were something better. They were skeletons held in a living hand to foretell fate. When Julius Caesar crossed the Rubicon, tipping history toward the end of the Roman Republic and the dawn of empire, his comment was *"Alea iacta est"*: The die is cast.

The Game of Dice, I suspect, will never end. These objects speak to something deep in our bones. Just remember to follow those six rules:

Good dice play fair.
Good dice look fair.
Good dice work well anywhere.
They're simple to roll.
They're hard to control.
And good dice feel nice.

Chapter 10

AN ORAL HISTORY OF THE DEATH STAR

MEMORIES OF THE GALAXY'S MOST FAMOUS SPHERE

Perhaps the greatest construction project in the history of geometry is the Death Star. Before being destroyed by a blond desert-boy in the tragic finale to the film *Star Wars*, it was pure terror. It was sheer beauty. It was a near-perfect sphere, a hundred miles across, armed with a planet-vaporizing laser. And yet even this behemoth, designed to compel the obedience of an entire galaxy, could not help but obey a higher master in turn: geometry.

Geometry yields to no one, not even evil empires.

I convened the team responsible for creating the Death Star to discuss the geometry behind history's most controversial solid. They brought up several considerations involved in building a tremendous spherical space station:

- Its manifold symmetries
- Its near-perpendicular surface relative to the direction of travel
- Its gravitational properties relative to naturally arising spheres
- Its personnel capacity as a function of surface area
- Its very slight curvature at large scale
- Its uniquely low surface-area-to-volume ratio

Even these powerful minds—architects, engineers, and grand moffs among them—could not dictate terms to geometry. Rather, they had to apply their ingenuity to work within its constraints. Here—in their own words—is how they did that.

Take note: For improved readability, I have edited out the loud, ominous breathing noises.

1. TO FRAME THY FEARFUL SYMMETRY

GRAND MOFF TARKIN

Our aim: To cow the galaxy with the greatest structure it had ever seen. Our funding: virtually infinite, thanks to the Emperor's unflinching approach to taxation. The sky was—quite literally—the limit. So our first question was: What should this thing look like?

IMPERIAL GEOMETER

I was told to find a simple, elemental design. Something so unnerving and spectacular that it would make droids weep and bounty hunters lose bladder control. Compared to peer-reviewing articles, it was a tough assignment.

GRAND MOFF TARKIN

That poor geometer spent a month brainstorming, filling notebooks with sketches and design ideas . . . all of which Lord Vader HATED.

DARTH VADER

A hexagonal prism? Was this to be an empire of honeybees?

IMPERIAL GEOMETER

Even in my deepest valleys of self-doubt, I was grateful for Lord Vader's feedback. Like a lot of visionaries, he is an exacting manager. But I know it's all meant as constructive criticism.

DARTH VADER

Imbecile.

What am I supposed to be, some kind of space pharaoh?

C'mon, we're the Empire, not the Borg.

You expect to conquer the galaxy in a hockey puck?

Ah, the Death Pencil. "Sketching out the next chapter of Galactic History." That's the least intimidating thing I've ever seen, and I've met Ewoks.

I'm sure Emperor Euclid will be pleased. Oh, wait — our emperor isn't a Greek geometer, he's a ruthless Machiavellian who kills people with dumb suggestions.

Now, this I can work with!

IMPERIAL GEOMETER

Eventually, we identified a single surpassing goal: symmetry.

Most people use the term casually, but in mathematics, "symmetry" has a precise meaning: it's a thing you can do to a shape that leaves it looking the same.

For example, a Wookiee face has a single symmetry: you can reflect it in a vertical mirror. That's it. If you do anything else—say, rotate it 90°, or reflect it in a horizontal mirror—then you rearrange the face entirely, after which the Wookiee may undertake to rearrange YOUR face.

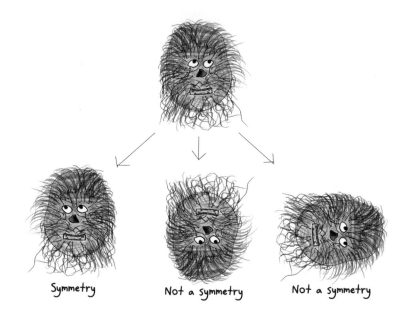

Symmetry | Not a symmetry | Not a symmetry

By contrast, the face of a dianoga—a tentacle-festooned swamp dweller, sometimes found in trash compactors—has three symmetries: two reflections, and a rotation of 180°.

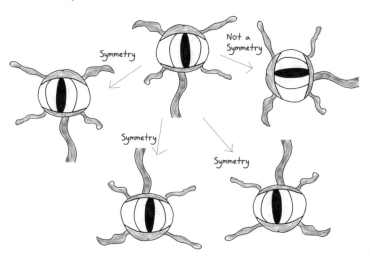

GRAND MOFF TARKIN

Why the obsession with symmetry? Well, symmetry is the essence of beauty.

Take faces. No one's is perfectly symmetrical. One ear is a little higher; one eye is a little larger; the nose is a bit crooked. But the more symmetrical the face, the more beautiful we find it. It's a strange fact of psychology: mathematics is the better part of beauty.

With the Death Star, we sought to create a force of awe and intimidation as powerful as the face of a supermodel.

IMPERIAL GEOMETER

One day, Lord Vader swept my illustrations from the table and roared, "MORE SYMMETRY!" We'd been considering an icosahedron, which has 120 symmetries. How could I top that? But then, in the proudest moment of my career—heck, of my LIFE—it came to me: the maximally symmetric shape.

DARTH VADER

Why did he not think of a sphere to begin with? One loses so much time to the glacial workings of feeble minds.

GRAND MOFF TARKIN

The headache wasn't over. We needed to install the planet-destroying laser array in the northern hemisphere, and thereby tarnish the symmetry. That ruffled some feathers.

IMPERIAL GEOMETER

I still maintain it was a mistake to include that weapon. I mean, which is more effective for intimidation: a little laser light show, or an INFINITE number of symmetries?

2. THROWING AERODYNAMICS
TO THE WIND

GRAND MOFF TARKIN

Right away, we encountered a problem. Ran into a headwind, you might say.

Back then, everybody was into Star Destroyers: these sleek, angular designs. They were galactic steak knives, ready to pop a star like a helium balloon. I thought the appeal was aesthetic, but when I learned the design was functional, too, it spelled trouble for our sphere.

IMPERIAL PHYSICIST

Imagine you're flying an airplane. No matter how good a pilot you are, you're going to have A LOT of collisions. I'm referring, of course, to air molecules.

Best-case scenario? The air molecules travel PARALLEL to your surface. Then, they won't impact you at all. They're like passing traffic in the neighboring lane. The worst-case scenario is that the air molecules hit PERPENDICULAR to your surface, at 90-degree angles. Then, your vessel bears the full force of the impact. That's why you don't build airplanes with big, flat fronts: it'd be like trying to weasel through a crowd while wearing a giant sandwich board on your torso.

Hence, the tapered design of the Star Destroyer. When it's traveling through atmosphere, the air molecules mostly glance off the sides—not quite parallel to the surface, but close enough. The Death Star, by contrast, gave us all aerodynamics nightmares. There's a huge surface hitting the air at near-perfect right angles.

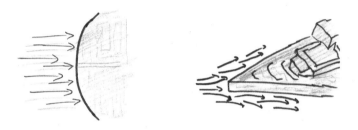

IMPERIAL ENGINEER

Imagine your friends are launching paper planes, and instead of joining in, you decide to throw a desk across the room. It's going to cost a lot more energy, and the flight won't be pretty.

GRAND MOFF TARKIN

Originally, we intended the Death Star to visit planets directly. It would enter the atmosphere, vaporize a continent or two, and blast "The Imperial March" through loudspeakers.

That dream died the moment we chose a sphere. Thwarted by aerodynamics, we were forced to keep our station up in the vacuum of space. No air resistance—but no music, either.

DARTH VADER

It was a harsh sacrifice. But leadership is no place for the fainthearted.

3. TOO BIG TO FAIL,
TOO SMALL TO SPHERE

GRAND MOFF TARKIN

Soon, another obstacle leapt into our path: our physicist kept insisting that the Death Star really ought to be lumpy and asteroid-shaped.

IMPERIAL PHYSICIST

Glance around the galaxy. Where do you see spheres? It's the big, heavy stuff. Stars, planets, a few larger moons. Now look at the smaller, less dense objects: asteroids, comets, dust clouds. You find a lot of weird potato shapes.

That's not a coincidence. It's a fact of gravity. I told them from the start: The Death Star is too small to become a sphere.

GRAND MOFF TARKIN

You should have seen Lord Vader's face.

DARTH VADER

I was about to break ground on the most ambitious construction project in the history of evil, and a lab coat with glasses was telling me it was TOO SMALL. Was I angry? You tell me.

IMPERIAL PHYSICIST

Look, I'm no propaganda expert, but the physics is pretty clear here. All matter attracts all other matter. More matter, more attraction. That's gravity.

So, toss a bunch of ingredients together in the mixing bowl of space, and every bit is mutually drawn toward every other bit. They congregate around a kind of 3D balancing point: the center of mass. Over time, the outlying clumps and more distant protrusions are drawn toward this center, until it reaches the final equilibrium shape: a perfect sphere.

But that's only if you've got enough matter. Otherwise, the attraction is too weak to conquer those lumpy outcroppings. Hence, big planets go spherical, but small moons stay potatoes.

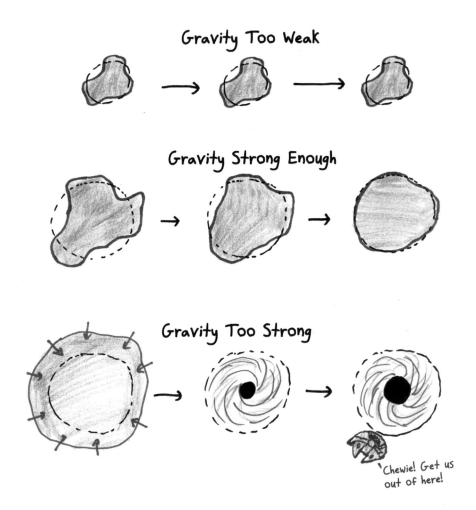

Gravity Too Weak

Gravity Strong Enough

Gravity Too Strong

Chewie! Get us out of here!

DARTH VADER

I wondered: Are all physicists this insolent? Should I drain the life out of every scientist right now, to spare myself future encounters?

IMPERIAL PHYSICIST

The magic size, where you're big enough to go spherical, depends on what you're made of. Ice will go spherical at a diameter of about 400 kilometers, because it's pretty malleable. Rock is much stiffer and needs more gravitational persuasion—so it won't go spherical until the diameter hits 600 kilometers. For a material like imperial steel, designed to withstand tectonic-level forces, it'd be even larger. Maybe 700 or 750 kilometers.

And the Death Star? It was only 140 kilometers across. A pebble.

GRAND MOFF TARKIN

Just as Lord Vader began to Force-choke the physicist, I was able to strike a compromise. "Gentlemen," I said, "this is a GOOD thing! It means that people seeing our artificial sphere will unconsciously assume that it's far bigger than it is. In effect, we'll triple our diameter without an extra dime for construction!" I also pointed out that we control the gravity anyway— as well as the physicist's air supply. At last, the physicist saw the wisdom.

IMPERIAL PHYSICIST

You could see Lord Vader's face light up. Well . . . I think you could. His mask looked smiley. Anyway, he clearly loved the idea that a spherical shape could imitate and evoke larger celestial bodies.

DARTH VADER

There is a fearsome sea creature that inflates itself into a sphere in order to intimidate its enemies. That is why, in my journals, I call the Death Star "the Pufferfish of the Skies."

GRAND MOFF TARKIN

I pleaded with him to stop using that "Pufferfish of the Skies" line, but you know how Lord Vader is. Once he latches on to an idea . . .

4. WEST VIRGINIA, FLOATING IN SPACE

GRAND MOFF TARKIN

Ever seen footage of the Death Star? You get the sense it was teeming with stormtroopers. Densely packed as a submarine.

Ha. In reality, the Death Star was the loneliest, emptiest place I've ever been.

IMPERIAL CENSUS TAKER

There were about 2.1 million people on the Death Star; that's counting droids. Meanwhile, with a radius of 70 kilometers, it had a surface area of almost 62,000 square kilometers. Now, assuming that you bring everybody to the surface level, you'll have a population density of about 30 people per square kilometer. That's five soccer fields per person.

To put that in perspective, it's about the same size, population, and population density as West Virginia. Want to picture social life on the Death Star? Imagine West Virginia floating in space.

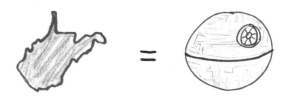

STORMTROOPER

Gosh, it got lonely sometimes. You'd be patrolling a sector all day with nobody in sight. Not even a droid.

IMPERIAL CENSUS TAKER

Of course, it was even worse than that. Not everybody spent their time on the surface! The habitable zones of the station went 4 kilometers down. At 4 meters per level, that's 1000 levels.

So, picture your floating Space West Virginia blanketed with coal mines, with people living up to 2.5 miles below the surface. On a given level, that's a population density of one person per 40 square kilometers. The only comparable place on Earth? Greenland.

STORMTROOPER

And the ultimate insult? They slept us 60 guys to a room. Each dorm had three toilets and two showers. Forget atomizing planets; if you want to give people nightmares, just show them the lines outside those bathrooms every morning.

The Emperor can afford a multi-quadrillion-dollar space orb, and here I am on board, sleeping in a triple-decker bunk bed and walking half a kilometer to pee? Ugh. It still makes me mad.

Place	Population Density (people per km²)
Crowded elevator	3,000,000
Submarine	50,000
New York City	10,000
Typical Suburb	1,000
State of Hawaii	90
Death Star (all on surface)	30
Death Star (all levels)	0.03
Mars during "The Martian"	0.000 000 007

5. CURVES YOU CAN'T SEE

GRAND MOFF TARKIN

Fundamentally, the Death Star wasn't about personnel. All the humans, all the machines—reactor core, sublight engine, hyperdrive—they were merely a support system for the station's real purpose: the superlaser.

IMPERIAL GEOMETER

Ah, that's another reason that a sphere was the perfect shape! For a given volume, it's got the lowest possible surface area. If you're building a casing for something huge—say, a planet-demolishing laser gun—then a sphere will get it done with the minimum of material.

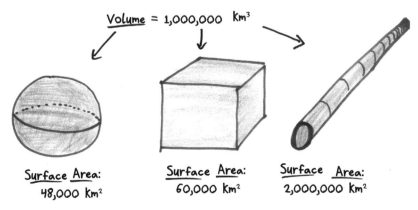

Volume = 1,000,000 km³

Surface Area: 48,000 km²

Surface Area: 60,000 km²

Surface Area: 2,000,000 km²

IMPERIAL ENGINEER

I know the geometer will tell you that the spherical shape saved us money, because a cube would have required 24 percent more steel. That's a typical mathematician for you: all theory, no practicality.

There's a reason we don't tend to build spherical ships: Curves are a pain! Ever tried to arrange furniture in a curved room? Good luck positioning the sofa.

Now, the Death Star was too big for you to notice the curvature on that level. No problem for the interior decorators. But on a construction level, that curvature gave me a multiyear migraine. You needed the steel beams to bend by about zero degrees, zero minutes, and 3 seconds of arc per meter —that's less than 1 degree of curvature per kilometer.

What a mess. The curvature was low enough that you couldn't see it with the naked eye, but high enough that every single part had to be custom-built.

GRAND MOFF TARKIN

The curved beams . . . Ugh, don't remind me. I remember one subcontractor shipped us straight beams for a whole year, thinking we wouldn't notice. Which, to be fair, we didn't, until one day the emperor is approaching by shuttle and says, "What's that funny bulge?" That set us back MONTHS. At least the subcontractor suffered worse than we did.

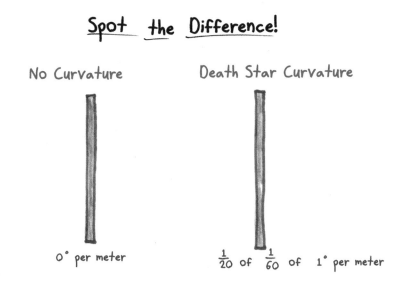

Spot the Difference!

No Curvature

Death Star Curvature

$0°$ per meter

$\frac{1}{20}$ of $\frac{1}{60}$ of $1°$ per meter

DARTH VADER

Foolish subcontractors. If you're going to stiff a customer, don't make it an evil empire.

6. MAYBE WE DID OUR JOBS
TOO WELL

IMPERIAL GEOMETER

Like I said, a sphere minimizes surface area. But I'll admit that it had other drawbacks.

GRAND MOFF TARKIN

Waste removal was a perpetual torment.

IMPERIAL SANITATION WORKER

The fundamental solution to trash in space is simple: jettison it. However, minimal surface area means the bulk of your station is far removed from the surface. We'd have needed trash chutes dozens of kilometers long. I launched a recycling campaign, but people just dumped everything—food, steel beams, live rebel prisoners—into the trash compactors. You couldn't help feeling that sustainability wasn't the priority.

IMPERIAL ENGINEER

The problem that still haunts me is the heating. It's outer space, right? Cold. You want to retain heat, and a sphere is great for that. Minimum surface area means minimum heat loss. But apparently, we did our jobs TOO well, because early simulations showed that the station would be prone to overheating.

IMPERIAL ARCHITECT

We had to dispose of some excess heat, so I put in thermal vents. Nothing big. A few meters wide. Release the heat into space; problem solved.

I didn't think . . .

I mean, when I heard that the rebels had destroyed the station by exploiting a thermal vent . . .

ATTORNEY FOR THE IMPERIAL ARCHITECT

Let the record show: The investigatory panel found that the Death Star was destroyed by a faulty reactor core, which my client did NOT design. My client's contribution was the thermal vents. These succeeded in their goal of transmitting heat out of the station.

DARTH VADER

They also succeeded in transmitting a proton torpedo INTO the station.

GRAND MOFF TARKIN

From shaky start to bitter end, the Death Star was a child of compromise. Was it expensive? Yes. Inefficient? No doubt. Destroyed by a ragtag band of rebel scum? Hard to deny.

And yet . . . I couldn't be prouder of that big, glorious sphere.

From a distance, it looked like a moon with an impact crater. Then you got closer, and the geometric perfection started to overwhelm you: the myriad symmetries of that crater, the right angles of the surface channels, the dark circular scar of the equatorial band . . .

IMPERIAL GEOMETER

The Death Star achieved this eerie fusion: so large it must have been natural, but so perfect it couldn't be. It was disturbing, compelling, horrifying. That's the power of geometry.

DARTH VADER

As often happens, our critics supplied our finest slogan. When my old frenemy Obi-Wan said, "That's no moon," we knew we had the tagline for the advertisements.

EUCLID ARCHIMEDES DYSON VADER PALPATINE

THE DEATH STAR

"That's no moon."

Obi-Wan Kenobi

DESTROYING YOUR PLANET SOMETIME SOON

$(1-\lambda)$ $2\pi\sqrt{a^2+b^2}$ $\quad |u \wedge v| = |u||v|\sin\theta$ $\qquad |AB| \cdot |BC| = \lambda \cdot (1-\lambda)$ $2\pi\sqrt{a^2+b^2}$ $\quad |u \wedge v| = |u||v|\sin\theta$

$b = \underline{b} - \underline{a} = \lambda(\underline{c} - a)$ $\quad (\underline{u} \cdot \underline{v})^2 + |u \wedge v|^2 = |\underline{u}|^2|v|^2$ $\qquad \vec{AC} = (\underline{c} - \underline{a}), \vec{AB} = \underline{b} - \underline{a} = \lambda(\underline{c} - a)$ $\quad (\underline{u} \cdot \underline{v})^2 + |u \wedge v|^2 =$

$\underline{c}$ $\quad \lambda = \frac{1}{2}$ $\qquad |u \wedge v|^2 = |u|^2|v|^2 \sin^2\theta$ $\qquad b = 11 - \lambda\}g + \lambda\underline{c}$ $\quad \lambda = \frac{1}{2}$ $\qquad |u \wedge v|^2 = |\underline{u}|^2|v|^2$

$u \wedge \underline{v} = |u||v|\sin\theta \underline{n}$ $\quad 2\pi\sqrt{a^2+b^2}$ $\quad b = \frac{1}{2}(g + \underline{c})$ $\qquad u \wedge \underline{v} = |u||v|\sin\theta$

$u \wedge v = A\underline{n}$
$A = |u||v|\sin\theta$
$t + \delta t$

$\underline{d} = \frac{1}{2}(b+c), \quad g = \frac{1}{3}(a+\underline{b}+\underline{c})$

$), \quad g = \frac{1}{3}(a+\underline{b}+\underline{c})$

$|u||v|\sin\theta$ area A

$\delta r = r(t+\delta t) \cdot r(t)$

$dt \to 0$

$|\underline{v}|^2 = |\underline{u}|^2|\underline{v}|^2$

$\frac{dr}{dt} = \lim_{\delta t \to 0} \frac{\delta r}{\delta t}$

$(\underline{u} \cdot \underline{v})^2 + |\underline{u} \wedge \underline{v}|^2 = |\underline{u}|^2|v|^2$

$U_3)^2 + (U_2 U_3 - U_5 U_2)^2 -$
$(U_4 V_2 - U_2 V_4)^2 =$

$(u_1 U_1 + u_2 v_2 + U_3 V_3)^2 + (U_2 U_3 - U_5 U_2)^2 -$
$(U_3 V_4 - U_4 V_3)^2 + (U_4 V_2 - U_2 V_4)^2 =$

$\frac{2}{3})(U_1^2 + V_2^2 + \frac{2}{3})$ $\qquad x^2 + y^2 = z^2$ $\qquad (\theta, \varphi) \in [0, 2\pi] \times$ $\qquad = (u_1^2 + u_2^2 + u_3^2)(U_1^2 + V_2^2 + \frac{2}{3})$ $\qquad x^2 + y^2 = z^2$

$[-\pi/2, \pi/2]$

$(1-\lambda)$ $2\pi\sqrt{a^2+b^2}$ $\quad |u \wedge v| = |u||v|\sin\theta$ $\qquad |AB| \cdot |BC| = \lambda \cdot (1-\lambda)$ $2\pi\sqrt{a^2+b^2}$ $\quad |u \wedge v| = |u||v|\sin\theta$

$b = \underline{b} - \underline{a} = \lambda(\underline{c} - a)$ $\quad (\underline{u} \cdot \underline{v})^2 + |u \wedge v|^2 = |\underline{u}|^2|v|^2$ $\qquad \vec{AC} = (\underline{c} - \underline{a}), \vec{AB} = \underline{b} - \underline{a} = \lambda(\underline{c} - a)$

$\underline{c}$ $\quad \lambda = \frac{1}{2}$ $\qquad |u \wedge v|^2 = |u|^2|v|^2 \sin^2\theta$ $\qquad b = 11 - \lambda\}g + \lambda\underline{c}$ $\quad \lambda = \frac{1}{2}$

$u \wedge \underline{v} = |u||v|\sin\theta \underline{n}$ $\quad 2\pi\sqrt{a^2+b^2}$ $\quad b = \frac{1}{2}(g + \underline{c})$

$u \wedge v = A\underline{n}$
$A = |u||v|\sin\theta$
$t + \delta t$

$\underline{d} = \frac{1}{2}(b+c), \quad g = \frac{1}{3}(a+\underline{b}+\underline{c})$

$), \quad g = \frac{1}{3}(a+\underline{b}+\underline{c})$

$|u||v|\sin\theta$ area A

$\delta r = r(t+\delta t) \cdot r(t)$

$dt \to 0$

$|\underline{v}|^2 = |\underline{u}|^2|\underline{v}|^2$

$\frac{dr}{dt} = \lim_{\delta t \to 0} \frac{\delta r}{\delta t}$

$(\underline{u} \cdot \underline{v})^2 + |\underline{u} \wedge \underline{v}|^2 = |\underline{u}|^2|v|^2$

$U_3)^2 + (U_2 U_3 - U_5 U_2)^2 -$
$(U_4 V_2 - U_2 V_4)^2 =$

$(u_1 U_1 + u_2 v_2 + U_3 V_3)^2 + (U_2 U_3 - U_5 U_2)^2 -$
$(U_3 V_4 - U_4 V_3)^2 + (U_4 V_2 - U_2 V_4)^2 =$

$\frac{2}{3})(U_1^2 + V_2^2 + \frac{2}{3})$ $\qquad x^2 + y^2 = z^2$ $\qquad (\theta, \varphi) \in [0, 2\pi] \times$ $\qquad = (u_1^2 + u_2^2 + u_3^2)(U_1^2 + V_2^2 + \frac{2}{3})$ $\qquad x^2 + y^2 = z^2$

$[-\pi/2, \pi/2]$

$(1-\lambda)$ $2\pi\sqrt{a^2+b^2}$ $\quad |u \wedge v| = |u||v|\sin\theta$ $\qquad |AB| \cdot |BC| = \lambda \cdot (1-\lambda)$ $2\pi\sqrt{a^2+b^2}$ $\quad |u \wedge v| = |u||v|\sin\theta$

$b = \underline{b} - \underline{a} = \lambda(\underline{c} - a)$ $\quad (\underline{u} \cdot \underline{v})^2 + |u \wedge v|^2 = |\underline{u}|^2|v|^2$ $\qquad \vec{AC} = (\underline{c} - \underline{a}), \vec{AB} = \underline{b} - \underline{a} = \lambda(\underline{c} - a)$

III

PROBABILITY
THE MATHEMATICS OF MAYBE

E ver flipped a coin? Unless you're too poor to have touched one, or too rich to have bothered, I'm going to guess "yes." I'll also guess that, even though the odds are 50/50, your most recent flip did not show a 50/50 mix of outcomes. It showed "heads" or "tails." All or nothing.

That's how life is: full of random one-off events. Unexpected train delays. Come-from-behind victories. Magical out-of-nowhere parking spots. In our storm-tossed world, anything can happen, and fate never RSVPs.

This world confuses the heck out of me.

SHOCKING NEWS

But if you could flip a trillion coins, you'd find yourself approaching a different world altogether: a well-groomed land of long-term averages. Here, half of all coins land on heads, half of all newborns are boys, and one-in-a-million events happen a millionth of the time, give or take. In this blue-skied theoretical realm, quirks and coincidences don't exist. They vanish against the totality of all possibilities, like stones thrown into the sea.

This world will never surprise you.

ALL IS USUAL

Probability bridges these two realms. The wild, woolly world we know is what makes probability necessary. The calm, measured world we can never quite reach is what makes probability possible. The probabilist is a dual citizen, aiming to understand every breaking headline and celebrity fiasco as a single card drawn from an infinite deck, a cup poured from a bottomless pitcher. As mortal creatures, we'll never set foot in the land of eternity—but probability can offer us a glimpse.

Chapter 11

THE 10 PEOPLE YOU MEET IN LINE FOR THE LOTTERY

A h, the lottery ticket. It's a certificate of optimism, a treasury bond from the Department of Hope. Why cling to a drab, dog-eared $1 bill when you could exchange it for a thrilling mystery bill worth anywhere from zero to $50 million?

If that sounds unappealing to you, well, you and humanity will just have to agree to disagree.

I should confess that I've spent less on lottery tickets in my life ($7) than I spent on croissants this month ($don't ask). Nevertheless, each year, roughly half of US adults play the lottery. It's not the half you might guess. People earning at least $90,000 are likelier to play than those earning below $36,000. Those with BAs participate more than those without. The state with the highest rate of lottery spending is my home state of Massachusetts: a haven of wealthy, overeducated liberals, who spend $800 per person each year on the lottery. Playing the lottery is like watching football, suing your neighbors, or butchering the national anthem: an American pastime, pursued for a variety of reasons.

Come, join me in line for a ticket, and let's investigate the multifaceted appeal of turning your money into commodified chance.

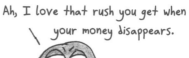

Ah, I love that rush you get when your money disappears.

1. THE GAMER

Behold! It's the Gamer, who buys lottery tickets for the same reason I buy croissants: not for sustenance but for pleasure.

Take the Massachusetts game entitled $10,000 Bonus Cash. It's a genius name. Put "10,000" and "bonus" in front of any word and there's no way you can go wrong. Plus, the graphics on the $1 tickets for this game look like rave night at the color printer. On the back, you'll find the following complicated odds of victory:

Prize	Odds
$ 10,000	1 in 1,000,800
$ 5,000	1 in 1,000,800
$ 500	1 in 50,400
$ 100	1 in 1,000
$ 40	1 in 1,007
$ 25	1 in 1,000
$ 20	1 in 300
$ 10	1 in 100
$ 5	1 in 150
$ 4	1 in 100
$ 3	1 in 100
$ 2	1 in 13.64
$ 1	1 in 10.71

So many exciting possibilities! It is truly the Buffet of Tasty Futures.

What is this ticket worth? Well, we don't know yet. Maybe $10,000; maybe $5; maybe (by which I mean "very probably") nothing.

It'd be nice to estimate its value with a single number. So imagine that we spent not a mere $1, but $1 million. This lets us escape the rowdy dance floor of the short-term world into the peace and quiet of the long-term one, where each payout arrives in the expected ratio. In our million tickets, a one-in-a-million event will occur roughly once. A 1-in-100,000 event will occur roughly 10 times. And a 1-in-4 event will occur 250,000 times, give or take.

Sifting through our stacks upon stacks of overstimulating paper, we'd expect the results to look something like this:

Prize	Tickets (per Million)
$ 10,000	1
$ 5,000	1
$ 500	20
$ 100	1000
$ 40	993
$ 25	1000
$ 20	3,333
$ 10	10,000
$ 5	6,667
$ 4	10,000
$ 3	10,000
$ 2	73,314
$ 1	93,371

Wow! 200,000 winners!
But wait... WAIT...

About 20% of our tickets are winners. Totaling up all of the prizes, our $1 million investment yields a return of roughly $700,000 . . . which means that we have poured $300,000 directly into the coffers of the Massachusetts state government.

Put another way: on average, these $1 tickets are worth about $0.70 each.

Mathematicians call this the ticket's *expected value*. I find that a funny name, because you shouldn't "expect" any given ticket to pay out $0.70, any more than you would "expect" a family to have 1.8 children. I prefer the phrase *long-run average*: It's what you'd make per ticket if you kept playing this lottery over and over and over and over and over . . .

Sure, it's $0.30 less than the price you paid, but entertainment ain't free, and the Gamer is happy to oblige. Poll Americans about why they buy lottery tickets, and half will say not "for the money" but "for the fun." These are Gamers. They're why, when states introduce new lottery games, overall sales rise. Gamers don't see the new tickets as competing investment opportunities (which would lead to a corresponding drop in sales for old tickets), but as fresh amusements, like extra movies at the multiplex.

What's the precise attraction for the Gamer? Is it the gratification of victory, the adrenaline rush of uncertainty, the pleasant buzz of watching it all unfold? Well, it depends on each Gamer's appetites.

I'll tell you what it isn't: the financial benefit. In the long run, lottery tickets will almost always cost more than they're worth.

Heh heh... I will soon be RICH! Or bankrupt. Whichever comes first.

2. THE EDUCATED FOOL

Wait . . . "Almost always"? Why "almost"? What state is dumb enough to sell lottery tickets whose average payout is more than their cost?

These exceptions emerge because of a common rule by which big-jackpot lotteries sweeten the deal: if no one wins the jackpot in a given week, then it "rolls over" to the next week, resulting in an even bigger top prize. Repeat this enough times, and the ticket's expected value may rise above its price. For example, in January 2016, the UK National Lottery draw offered a juicy expected value of more than £4, compared to a ticket price of only £2. Weird as they appear, such schemes typically spur more than enough sales to justify their cost.

In the queue for lotteries like this, you'll meet a very special player, a treat for gambling ornithologists like us. See it there, high in the branches, grooming itself? It's the Educated Fool, a rare sap-brained creature who does with "expected value" what the foolish always do with education: mistake partial truth for total wisdom.

"Expected value" distills the multifaceted lottery ticket, with all its prizes and probabilities, down to a one-number summary. That's a powerful move. It's also simplistic.

Take these two tickets, each costing $1.

Price: $1 **Ticket A**
Prize: $ 9 million
Odds: 1 in 10 million

Hideous! Overpriced!
The mere suggestion
offends me!

Price: $1 **Ticket B**
Prize: $ 11 million
Odds: 1 in 10 million

Glorious! A steal!
I'll take five thousand!

Spend $10 million on A, and you expect to earn back only $9 million, amounting to a $0.10 loss per ticket. Meanwhile, $10 million spent on B should yield $11 million in prizes, and thus a profit of $0.10 per ticket. So to those enamored with expected value, the latter is a golden opportunity, while the former is pyrite trickery.

And yet . . . would $11 million bring me any greater happiness than $9 million? Both values exceed my current bank account many times over. The psychological difference is negligible. So why judge one a rip-off and the other a sweet bargain?

Even simpler, imagine that Bill Gates offers you a wager: for $1, he'll give you a 1 in a billion chance at $10 billion. Calculating the expected value, you start to salivate: $1 billion spent on tickets would yield an expected return of $10 billion. Irresistible!

Even so, Educated Fool, I beg you to resist. You can't afford this game. Scrape together an impressive $1 million, and the 99.9% likelihood is still that rich man Gates walks away $1 million richer while you walk away broke. Expected value is a long-run average, and with Gates's offer, you'll exhaust your finances well before the "long run" ever arrives.

The same holds true in most lotteries. Perhaps the ultimate repudiation of expected value is the abstract possibility of $1 tickets like this:

If you buy 10 tickets, you're likely to win $1. That's pretty terrible: only $0.10 per ticket.

If you buy 100 tickets, you're likely to win $20. (That's 10 of the smallest prize, and one of the next smallest.) Slightly less terrible: now $0.20 per ticket.

If you buy 1000 tickets, you're likely to win $300. (That's a hundred $1 prizes, ten $10 prizes, and a single $100 prize.) We're up to $0.30 per ticket.

Keep going. The more tickets you buy, the better you can expect to do. If you somehow buy a trillion tickets, the likeliest outcome is that you will win $1.20 per ticket. A quadrillion tickets? Even better: $1.50 per ticket. In fact, the more tickets you buy, the greater your profit per ticket. If you could somehow invest $1 googol dollars, you'd be rewarded with $10 googol in return. With enough tickets, you can attain any average return you desire. The ticket's expected value is infinite.

But even if you trusted a government to pay out, you could never afford enough tickets to glimpse that profit. Go ahead and spend your life savings on these, Educated Fool. The overwhelming likelihood is bankruptcy.

We humans are short-run creatures. Better leave the long-run average to the immortals.

Who, me? I'm just the hired gun.

3. THE LACKEY

Ooh, look who just stepped into the queue. It's the Lackey!

Unlike most folks here, the Lackey's windfall is guaranteed. That's because the Lackey isn't keeping the tickets, just pulling down minimum wage to stand in line, purchase them, and pass them off to someone else. Modest compensation, but ultrareliable.

Who would pay such a Lackey? Well, that's a question for our next player . . .

Eeeeexcellent. All is going according to plan.

4. THE BIG ROLLER

At first glance, this character looks an awful lot like the Educated Fool: the same gleaming eye, the same scheming grin, the same obsessive focus on expected value. But watch what happens when he encounters a lottery with positive expected value. Whereas the Fool flails, buying a hapless handful of tickets and rarely winning, the Big Roller masterminds a simple and nefarious plan. To transcend risk, you can't just buy a few tickets. You've got to buy them all.

How to become a Big Roller? Just follow these four steps, as elegant as they are loony.

Step #1: **Seek out lotteries with positive expected value.** This isn't as rare as you'd think. Researchers estimate that 11% of lottery drawings fit the bill.

Step #2: **Beware the possibility of multiple winners.** Big jackpots draw more players, increasing the chance that you'd have to split the top prize. That's an expected-value-diminishing disaster.

Step #3: **Keep an eye on smaller prizes.** In isolation, these consolation prizes (e.g., for matching four out of six numbers) aren't worth much, but their reliability makes them a valuable hedge. If the jackpot gets split, small prizes can keep the Big Roller from losing too much.

Finally, *Step #4*: **When a lottery looks promising, buy every possible combination.**

Sounds easy? It's not. The Big Roller needs extraordinary resources: millions of dollars of capital, hundreds of hours to fill out purchase slips, dozens of Lackeys to complete the purchases, and retail outlets willing to cater to massive orders.

To understand the challenge, witness the dramatic tale of the 1992 Virginia State Lottery.

That February's drawing offered the perfect storm of circumstances. Rollovers had driven the jackpot up to a record $27 million. With only 7 million possible number combinations, that gave each $1 ticket an expected value of nearly $4. Even better, the risk of a split prize was reassuringly small: only 6% of previous Virginia lotteries had resulted in shared jackpots, and this one had climbed so high that it would yield profit even if shared three ways.

And so the Big Roller pounced. An Australian syndicate of 2500 investors, led by mathematician Stefan Mandel, made their play. They worked the phones, placing enormous orders at the headquarters of grocery and convenience store chains.

The clock labored against them. It takes time to print tickets. One grocery chain had to refund the syndicate $600,000 for orders it failed to execute. By the time of the drawing, the investors had acquired only 5 million of the 7 million combinations, leaving a nearly 1 in 3 chance that they would miss out on the jackpot altogether.

Luckily, they didn't—although it took them a few weeks to surface the ticket from among the stacks of 5 million. After a brief legal waltz with the state lottery commissioner (who vowed "This will not happen again"), they claimed their prize.

Big Rollers like Mandel enjoyed profitable glory days in the 1980s and early 1990s. But those heady times have vanished. As tricky as the Virginia buyout was, it was a logistical breeze compared to the daunting prospect of buying out Mega Millions or Powerball (each with more than 250 million possible combinations). Factor in post-Virginia rules designed to thwart bulk ticket purchases, and it seems the Big Roller may never find the right conditions for a return.

Oh, don't mind me.
I'm just here because
I find you all fascinating!

5. THE BEHAVIORAL ECONOMIST

For psychologists, economists, probabilists, and various other *-ists* of the university, nothing is more intriguing than how people reckon with uncertainty. How do they weigh danger against reward? Why do some risks appeal while others repel? But in addressing these questions, researchers encounter a problem: Life is super complicated. Ordering dessert, changing jobs, marrying that good-looking person with the ring—to make these choices is to roll an enormous die with an unknown number of irregular faces. It's impossible to imagine all the outcomes or to control for all the factors.

Lotteries, by contrast, are simple. Plain outcomes. Clear probabilities. A social scientist's dream. You see, the behavioral economist is not here to play, just to watch *you* play.

The scholarly fondness for lotteries goes back centuries. Take the beginnings of probability in the late 1600s. This age saw the birth of finance, with insurance plans and investment opportunities beginning to spread. But the nascent mathematics of uncertainty didn't know what to make of those complex instruments. Instead, amateur probabilists turned their eye to the lottery, whose simplicity made it the perfect place to hone their theories.

More recently, the wonder duo of Daniel Kahneman and Amos Tversky identified a powerful psychological pattern on display in lotteries. For that, I introduce you to our next line-buddy . . .

6. THE PERSON WITH NOTHING LEFT TO LOSE

In the spirit of behavioral economics, here's a fun would-you-rather:

#1: Which option would you prefer?

A: Guaranteed $900

B: 90% chance at $1000

Hmm... take the guarantee, or risk it all for a little extra?

In the long run, it doesn't matter. Choose B 100 times, and (on average) you'll get 90 windfalls, along with 10 disappointing zeroes. That's a total of $90,000 across 100 trials, for an average of $900. Thus, it shares the same expected value as A.

Yet if you're like most people, you've got a strong preference. You'll take the guarantee, rather than grab for a little extra and risk leaving empty-handed. Such behavior is called **risk-averse**.

Now, try this one:

Hmm... suffer a certain loss, or risk a greater one for the chance to pay nothing?

#2: Which option would you prefer?

A: Guaranteed loss of $900

B: 90% chance of losing $1000

It's the mirror image of Question #1. In the long run, each yields an average loss of $900. But this time, most people find the guarantee *less* appealing. They'd rather accept a slight extra penalty in exchange for the chance to walk away scot-free. Here, they are **risk-seeking**.

These choices are characteristic of *prospect theory*, a model of human behavior. When it comes to gains, we are risk-averse, preferring to lock in a guaranteed profit. But when it comes to losses, we are risk-seeking, willing to roll the dice for the chance to avoid a bad outcome.

A crucial lesson of prospect theory is that framing matters. Whether you call something a "loss" or a "gain" depends on your current reference point. Take this contrast:

I give you $1000. Which do you prefer?

A: Another $500 guaranteed

B: 50% chance at another $1000

I give you $2000. Which do you prefer?

A: lose $500 of it

B: 50% chance of losing $1000 of it

The questions offer identical choices: (a) you walk away with $1500, or (b) you flip a coin to determine whether you end with $1000 or $2000. But folks don't give identical responses. In the first case, they prefer the $1500 guaranteed; in the second, they favor the risk. That's because the questions create different reference points. When the $2000 is "already yours," the thought of losing it stresses you out. You're willing to take risks to avoid that fate.

When life is hard, we roll the dice.

This line of research has a sad and illuminating implication for lottery ticket purchasers. If you're living under dire financial conditions—if every day feels like a perpetual loss—then you're more willing to risk money on a lottery ticket.

Think of how a basketball team, trailing late in the game, will begin fouling its opponents. Or how a hockey team, down a goal with a minute to play, replaces its goalie with an extra forward. Or how a political candidate, behind with two weeks until the election, will go on the attack, hoping to shake up the campaign. These ploys harm your expected value. You'll probably lose by an even larger margin than before. But by heightening the randomness, you boost your chances at winning. In times of desperation, that's all people want.

Researchers find that those who buy lottery tickets "for the money" are far likelier to be poor. For them, the lottery isn't an amusing way to exercise wealth, but a risky path to acquire it. Yes, on average, you lose. But if you're losing already, that's a price you may be willing to pay.

Yes! I can vote, smoke, gamble, and enlist in the military. Tonight is going to be a fun night.

7. THE KID WHO JUST TURNED 18

Hey, look at that fresh-faced young lottery player! Does the sight fill you with nostalgia for your own youth?

Or does it make you think, "Hey, why are states in the business of selling a product so addictive that they feel compelled to protect minors from buying it?"

Well, there's someone else I'd like you to meet . . .

I don't buy this ticket for me. I buy it for America.

8. THE DUTIFUL TAXPAYER

Say hello to our next friend, a would-be civic hero: the Dutiful Taxpayer.

Nobody likes being hit up for money, even by the friendly neighborhood government. States have tried taxing all sorts of things (earnings, retail, real estate, gifts, inheritance, cigarettes), and none of it brings their citizens any special joy. Then, in the 1970s and 1980s, these governments stumbled upon an ancient mind trick.

Turn tax paying into a game, and people will line up down the street to play.

The proposition "I'll keep \$0.30 from each dollar" prompts grumbles. Adjust it to "I'll keep \$0.30 from each dollar and give the entire remainder to one person chosen at random," and you've launched a phenomenon. States don't offer lotteries out of benevolence; they do it to exploit one of the most effective moneymaking schemes ever devised. State lotteries deliver a profit of \$30 billion per year, accounting for more than 3% of state budgets.

If you're picking up a funny odor, I suspect it's hypocrisy. It's a bold move to prohibit commercial gambling and then run your own gambling ring out of convenience stores. Heck, the "Daily Numbers" game that states embraced during the 1980s was a deliberate rip-off of a popular illegal game.

Is there an innocent explanation? To be fair, public lotteries are less corruption-prone than private ones. Just ask post-USSR Russians, who found themselves inundated with unregulated mob-run lotteries. Still, if we're merely aiming to shield gamblers from exploitation, why run aggressive advertising campaigns? Why pay out so little in winnings? And why print gaudy tickets that resemble nightclub fliers spliced with Monster Energy drinks? The explanation is obvious: lotteries are for revenue, period.

To fend off attacks like the one I'm making now, states employ a clever ruse: earmarking the funds for specific popular causes. "We're running a lottery to raise money" sounds better when you append "for college scholarships" or "for state parks" than it does with "for us—you know, to spend!" This tradition goes way back: Lotteries helped to fund churches in 15th-century Belgium, universities like Harvard and Columbia, and even the Continental Army during the American Revolution.

Earmarks soften the lottery's image, helping to attract otherwise nongambling citizens. These are your Dutiful Taxpayers: not necessarily inclined toward games of chance, but willing to play "for a good cause." Alas, they're getting hoodwinked. Tax dollars are fungible, and lottery revenue is generally offset, dollar for dollar, by cuts in spending from other sources. As comedian John Oliver quips: designating lottery money for a particular cause is like promising "to piss in one corner of a swimming pool. It's going all over the place, no matter what you claim."

So why do earmarks—and state lotteries, for that matter—endure? Because a play-if-you-feel-like-it lottery sounds a lot better than a pay-or-you'll-go-to-jail tax.

9. THE DREAMER

Now, allow me to introduce you to a starry-eyed hopeful, who holds as special a place in my heart as the lottery holds in theirs: the Dreamer.

For the Dreamer, a lottery ticket isn't a chance to win money. It's a chance to fantasize about winning money. With a lottery ticket in hand, your imagination can go romping through a future of wealth and glory, champagne and caviar, stadium skyboxes and funny-shaped two-seater cars. Never mind that winning a lottery jackpot tends to make people less happy, a trend well documented by psychologists. Daydreaming about that jackpot affords a blissful few minutes while driving your disappointing regular-shaped car.

The prime rule of fantasy is that the top prize must be enough to change your life, transporting you into the next socioeconomic stratum. That's why instant games, with their modest top prizes, draw in low-income players. If you're barely scraping together grocery money each week, then $10,000 carries the promise of financial transformation. By contrast, comfortable middle-class earners prefer games with multimillion-dollar jackpots—enough kindling for a proper daydream.

If you're seeking an investment opportunity, it's crazy to fixate on the possible payout while ignoring the probability. But if you're seeking a license for fantasy, then it makes perfect sense.

This Dreamer tendency helps explain the lottery's odd economy of scale, whereby big states fare better than small ones. Imagine a simplified lottery game: Half of all revenue goes to the state, and the other half goes to the jackpot winner. If a state sells $1 million in tickets, then each player has a 1-in-a-million chance at a $500,000 jackpot. Meanwhile, if a state can sell $100 million in tickets, then the odds drop to 1 in 100 million, and the jackpot leaps to $50 million.

LIL' STATE LOTTO

Prize: $ 500,000

Odds: 1 in 1,000,000

Half a mill.
Decent fantasy.

BIG OL' STATE LOTTO

Prize: $ 50,000,000

Odds: 1 in 100,000,00

Fifty million! Oh my word, this is top-shelf fantasy kindling.

The expected value doesn't change. In each state, an average ticket is worth $0.50. But the psychology of the Dreamer will favor the megajackpot with the miniscule odds.

Aha, I win again!
Best game ever!

10. THE ENTHUSIAST FOR SCRATCHING THINGS

Ah, this citizen knows where it's at. Forget the cash prizes and all that probability mumbo-jumbo. The chance to rub a quarter across some cardboard is winnings enough. The state lottery: it's like a scratch-and-sniff for grown-ups.

Chapter 12

CHILDREN OF THE COIN

I f I've been stumped and impressed by one thing in my teaching career, it's human genetics. I'm not referring to that unfortunate semester in 2010 when I was pressed into teaching 10th-grade biology. Rather, I refer to the job's finest long-term perk: getting to know families.

Every time you teach a pair of relatives—brother and sister, cousin and cousin, aunt and equal-aged nephew—you renew your wonder at the scattershot nature of biological inheritance. I've taught siblings who looked like twins, and twins who looked like strangers. Parents' nights always mess with my head. Meeting after meeting, my mind performs a real-time face-mash of the two adults before me, and I discover that their kid is a seamless Photoshop job: his ears flanking her eyes, his hair on her head shape. All families are alike, and yet no two families are alike in quite the same way.

The riddles of resemblance strike at the heart of biology. But as my students can tell you, I am not a biologist. I have no DNA sequencer, no special knowledge of introns or histones, and (more to the point) no clue. No, I'm a mathematician. What I've got is a coin, a theorem, and a *Brave Little Toaster*–esque faith in the power of figuring things out from first principles.

And maybe that's enough.

This chapter is about two seemingly unrelated questions in the field of probability. First is the question of genetic inheritance, which could fill grad-school textbooks. Second is the question of coin flips, which feels so trivial it's barely worth stooping to pick up off the ground.

Can we bridge the two realms? Can the coin's monosyllabic grunts give expression to the complexity of the human race?

I want to say yes. And I'm pretty sure I got that from my dad.

Okay, let's start with the easier of the two questions: What happens when you flip a coin?

Answer: The two outcomes, heads and tails, occur with equal likelihood. Solved!

Mmm, you feel that? It's the pleasure of turning away from an urgent, intractable real-world problem to focus instead on a puzzle that nobody cares about. Savor the sensation. It's why people become mathematicians.

Okay, so nothing much happens when we toss a single coin. But what if we toss a pair? You'll find that four outcomes are equally likely:

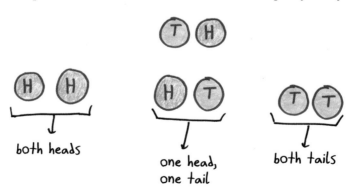

A stickler for specifics will see the two middle outcomes as different: head then tail vs. tail then head. (Imagine that one coin is a penny and the other is a nickel.) But if you're like most coin flippers, you won't care. Your mind collapses both into a single outcome ("one head, one tail") that happens to have double the probability.

Keep going: What if we toss three coins?

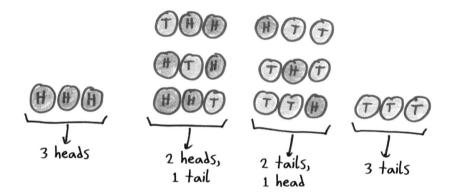

Eight outcomes in all. If you're hoping for all heads, there's only one path: every coin must perform its duty, so the probability is 1 in 8. But if you're aiming for two heads and a tail, then you've got three ways to do it—the tail can appear on the first, second, or third coin—giving a probability of 3 in 8.

Now, what about four coins—or five, or seven, or 90?

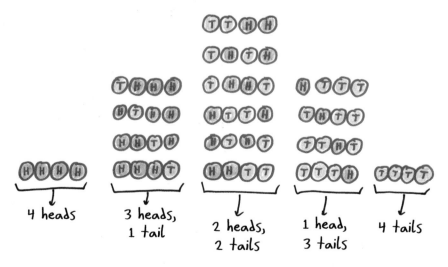

What we're exploring here is a family of probabilistic models known as the *binomial distribution*. You take a single binary event (heads or tails, win or lose, one or zero) and repeat it several times. How many of each outcome will occur? That's the binomial. And, pursuing a branch of this mathematical family, we witness two clear trends.

First, as we add coins, the complexity blooms. An extra coin doubles the number of possible results, from two to four to eight to 16. Yet the orchestra is only tuning up. Make it 10 coins, and we'll exceed 1000 possibilities. Reach 50 coins, and we'll face 2^{50} possible outcomes: more than a quadrillion. This aggressive growth pattern is known as the *combinatorial explosion*. A linear increase in the number of objects leads to an exponential increase in the number of ways to combine them. A handful of coins, simple as it seems, gives rise to unfathomable variety.

The second trend points the other direction: Even as the variety explodes, the extremes vanish. Check out the four-coin case above, which has only two "extreme" outcomes (all heads or all tails) against 14 "middle" outcomes (with some of each). The more coins we toss, the less likely the extremes become and the more we find ourselves mired in the swampy middle ground of "a pretty even mix of heads and tails."

Number of Coins	Total Possible Outcomes	Likelihood of "Extreme" Outcome (more than 75% or less than 25% heads)
1	2	100%
5	32	37.5%
10	1024	10.9%
20	1 million	4.1%
30	1 billion	0.5%
40	1 trillion	0.2%
100	1000 million billion trillion	0.00006%

The reason is simple. Ordinary outcomes are ordinary precisely because there are so many ways to achieve them. Meanwhile, extreme outcomes are extreme precisely because they are so singular, unfolding in very few ways.

Suppose we flip 46 coins. There is only one way to achieve all heads.

But allow yourself just one tail among 45 heads, and the possibilities multiply: the tail could come first, or second, or third, or fourth, or fifth . . . all the way up to the tail coming last. That's 46 combinations.

Make it two tails among 44 heads, and it's even easier to achieve. The tails could come first and second, or first and third, or first and fourth . . . or first and last . . . or second and third, or second and fourth, or second and fifth . . . all the way up to last and second to last. That's more than 1000 possibilities in total, and we're just getting warm. The likeliest option—23 tails and 23 heads—can occur in 8 trillion distinct ways.

I'm no psychic, but if you go flip 46 coins, I predict that the following will happen:

1. You will get between 16 and 30 heads.
2. You will flip a specific sequence that no 46-coin flipper in history has ever flipped: a one-time historical anomaly, unique in human experience.

Somehow, though, the first point (your outcome is middling) will overshadow the second (your outcome is glorious and unique). To us, coins are interchangeable, so any roughly balanced sequence will strike us as generic and forgettable. A snowflake in a blizzard.

But imagine if we cared about the specific constitution of that snowflake. Imagine if each heads and tails struck us as a potent twist of fate. What if we worried about not just how many heads came up but which ones precisely?

Heads heads tails?! Extraordinary! Distinct! Irreplaceable!

Then, whichever of the 70 trillion possible combinations emerged would strike us as so vanishingly unlikely, so egregiously improbable, that its appearance would seem a miracle. It would enchant us like a star plucked from the sky and lowered into our hands. Those 46 coins would be as precious as . . . well, as a newborn baby.

And that brings us to the chapter's hard question: genetics.

Every cell in your body contains 23 pairs of chromosomes. Think of them as a 23-volume recipe book, the instructions for building you. You have two versions of each volume: one from your mother, and one from your father.

Of course, your mother and father each have two copies themselves: one from your grandmother, and one from your grandfather. How did they decide which one to pass on to you, their precious little pea plant? Well—and here I simplify rather dramatically—they flipped a coin. Heads: you get Grandpa's chromosome. Tails: you get Grandma's.

Your mother and father repeated this process 23 times each, once per volume, and the result was . . . well, you.

By this model, each couple can generate 2^{46} distinct offspring. That's 70 trillion. And unlike with coins, the specifics matter. I have my mother's thick hair and pleasure in reading, as well as my father's gait and his love of clarity. If I had a different mix of their traits—my father's curls, say, and my mother's height (or lack thereof)—then I'd be a different person, a sibling to myself.

In genetic inheritance, a head and a tail is not the same as a tail and a head.

This model predicts the varying degrees of resemblance we see between siblings. At one extreme, every coin flip for the younger sibling could come up exactly as it did for the older one. Such siblings, despite being separated by years, would be, in essence, identical twins.

At the other extreme, it is possible for *none* of the coin flips to come up the same. Such siblings, as in a creepy Philip K. Dick story, would be genetic strangers, sharing no closer a biological relation than did the parents who birthed them.

Of course, both extremes are wildly improbable. The far greater likelihood is that two siblings will share roughly half of their 46 chromosomes in common—perhaps a little more, perhaps a little less. According to our binomial model, the vast majority will share between 18 and 28.

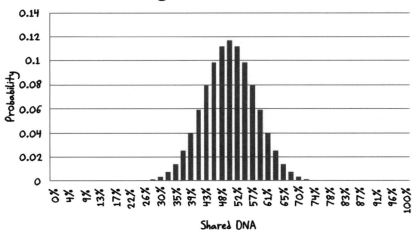

In fact, data on nearly 800 sibling pairs, compiled by Blaine Bettinger at his blog *The Genetic Genealogist*, matches our crude prediction pretty darn well:

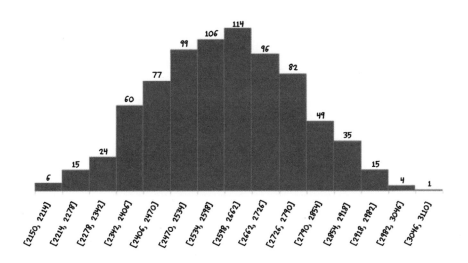

There's a caveat. I've omitted a factor so important that I can't blame biologists if they're rage-tearing the pages of this book right now. That's *crossing over*, or *recombination*.

I've claimed that chromosomes are passed down undisturbed and intact. Like many things I said while teaching biology, that's false. Before a single version of each chromosome is selected, the two experience a splicing-together—for example, swapping the middle third of each. Thus, a given chromosome may come *mostly* from Grandpa, but it'll feature a handful of recipes exchanged for the equivalents from Grandma.

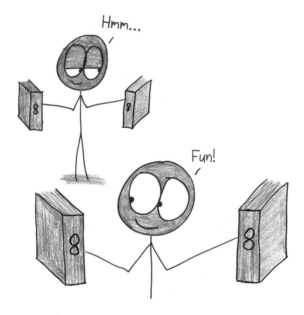

Crossing over happens roughly twice per chromosome. Thus, to enhance the accuracy of our model, we can triple the number of coin flips (since two crossovers will split a chromosome into three pieces).

How does this impact the number of possible offspring? Remember, linear growth in the number of objects yields exponential growth in their combinations. So the variety of offspring does far more than triple. In fact, it grows from 2^{46} (or 70 trillion) to 2^{138} (which is a staggering 350 duodecillion).

I'm arguing, in short, that a newborn baby has a lot in common with a handful of scattered pennies. That's not to say you should revalue your infant's life at $0.46. To the contrary, you should admire $0.46 of spilled pocket change as a miracle on par with human birth.

Getting to know families, you feel that they mix like liquids, blue and yellow paint sloshing together to create green. But they don't. Families mix like decks of cards, with elements shuffled and reshuffled, tucked up one's sleeve and then brought back into play. Genetics is a game of combinations: the lovely and elusive patterns of resemblance between siblings can all be traced back to the combinatorial explosion. Flip enough coins, and the stark, distinct outcomes (heads vs. tails) begin to blur and blend. The jagged edges of the graph fuse to form something as liquid and flowing as humanity itself. In this way, we are the offspring of combinatorics, the progeny of a shuffled deck, the children of the coin.

Chapter 13

WHAT DOES PROBABILITY MEAN IN YOUR PROFESSION?

o call human beings "bad" at probability would be simplistic and
mean-spirited. Probability is a subtle branch of modern mathematics, booby-trapped with paradoxes. Even an elementary problem
can throw a stone-cold expert for a loop. Dissing folks for their probabilistic failures is a bit like calling them bad at flying, or subpar at swallowing
oceans, or insufficiently fireproof.

No, it would be fairer to say that human beings are *terrible* at probability.

In their psychological research, Daniel Kahneman and Amos Tversky found that people harbor persistent misconceptions about uncertain
events. Time and again, they overrate negligible possibilities, and under-
rate near certainties.

Actual Probability	How People Treat It
1%	6%
5%	13%
10%	19%
50%	42%
90%	71%
95%	79%
99%	91%

1%, eh? So you're saying there's a chance... ...like, 1 in 20 or so...

I want a near-guarantee! Like, 99 in a hundred! Not this 99% nonsense.

No big deal, right? I mean, does probability ever come up in the real world? It's not like we spend our lives clawing for intellectual tools that might offer the slightest stability in the swirling miasma of uncertainty that surrounds us every waking moment . . .

Well, just in case—this chapter is a handy guide on how various sorts of human beings think about uncertainty. Just because it's hard doesn't mean we can't have some fun with it.

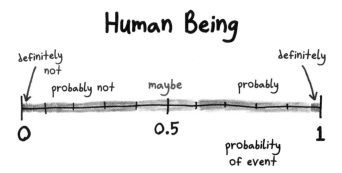

Hello! You are a person. You have two eyes, a nose, and sentience. You dream, laugh, and urinate, not necessarily in that order.

Also, you live in a world where nothing is certain.

Take a simple fact: How many planets does your solar system have? These days, you say "eight." From 1930 until early 2006, you answered "nine" (counting Pluto). In the 1850s, you wrote textbooks listing twelve: Mercury, Venus, Earth, Mars, Jupiter, Saturn, Uranus, Neptune, Ceres, Pallas, Juno, and Vesta. (You now call the last four "asteroids.") In 360 BCE, you named seven: Mercury, Venus, Mars, Jupiter, Saturn, the moon, and the sun. (You now call the last two "the moon" and "the sun.")

You keep changing your mind as new data and new theories emerge. That's classic you: When it comes to knowledge, you've got lots of good ideas but no guarantees. Teachers, scientists, politicians, even your sense organs—any of these could deceive you, and you know it.

Probability is the language of your uncertainty. It lets you quantify what you know, what you doubt, what you doubt you know, and what you know you doubt, expressing these nuances of confidence in a clear and quantitative language. At least, that's the idea . . .

Political Journalist

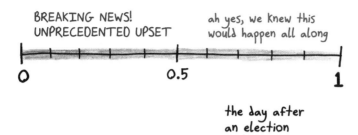

Hello! You are a political journalist. You write about elections coming up. You write about elections going down. On rare special days, you even write about things like "policy" and "governance."

Also, you seem baffled by the occurrence of slightly unlikely events.

This wasn't always true. Once upon a time, you saw elections as magical moments of infinite possibility. You downplayed the likeliest outcome so as to build suspense, to make it seem as if each race was won on a half-court buzzer-beater. On election night 2004, when George W. Bush led Ohio by 100,000 votes with barely 100,000 left to count, you said the state was "too close to call." When probabilistic models gave Barack Obama a 90% chance to win in 2012, you dubbed the race "a toss-up."

Then, 2016 turned your world to pineapple upside-down cake. Donald Trump defeated Hillary Clinton. Waking up the next day, you felt you'd experienced a quantum singularity, no more foreseeable than a rabbit materializing out of thin air. But to probabilist Nate Silver and like-minded others, this was only a moderate surprise, a 1-in-3 event—like rolling a die that lands on 5 or 6.

Investment Banker

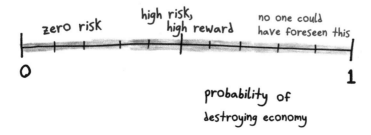

Hello! You are an investment banker. You bank investments. You invest in banks. And your suits are worth more than my car.

As recently as the 1970s, your job was pretty boring. You were a funnel for funds, pouring money into "stocks" (i.e., pieces of companies) or "bonds" (i.e., pieces of debt). Stocks were exciting; bonds were bland. You were a human bond.

Now, your job is as thrilling as a roller coaster that has failed its safety inspection. Starting in the 1970s and 1980s, you began inventing complex financial instruments that no one fully understands, least of all the government regulators. Sometimes, these devices earn you big bonuses. Sometimes, they make your century-old firm collapse, leaving a vast crater across the economy as if a dinosaur-killing asteroid has struck. Exciting times!

To be fair, allocating capital is a pretty important job in capitalism, and has the potential to create lots of value. It annoys you when snarky math teachers slight your profession. If so, take a moment to calculate the factor by which your salary exceeds that of the snarky math teachers, and I think you'll find your spirits rising.

Local News Anchor

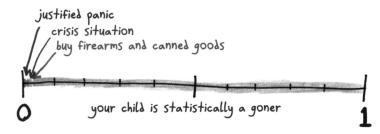

Hello! You are a local news anchor. You have great hair, crisp enunciation, and a long history of artificial banter with your coanchor.

Also, you are obsessed with very unlikely events.

You like to report on chilling dangers. Local murders. Airborne carcinogens. Defective toys that attach themselves like the creature in *Alien* to your baby's chubby cheeks. Ostensibly, you do this to inform your audience. But be honest. You do it to hold their attention. If you hoped to raise awareness about the statistical dangers to children, you'd warn about household accidents caused by guns and swimming pools. Instead, you paint vivid landscapes in which the only danger more pressing than abduction is crocodile attacks. You know that we can't tear our eyes away from such horrors—especially when they're leavened with artificial banter.

Weather Forecaster

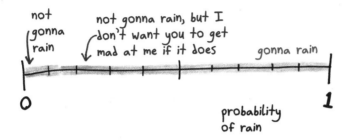

Hello! You are a weather forecaster. A televised cloud prophet. You gesture with conviction, and end every conversation with "And now, back to you."

Also, you fudge probabilities so people don't get mad at you.

Of course, you try to be honest. When you say that there's an 80% chance of rain, you're exactly right: It rains on 80% of such days. But when rain is less likely, you inflate the numbers. You fear angry tweets from people who leave their umbrellas at home and then blame you for the sky's damp doings. So when you forecast a 20% chance of rain, it rains only 10% of the time. You're increasing the odds so as to decrease your hate mail.

Perhaps, if people understood probability better, you could speak your truth. People seem to act as though "10%" means "not going to happen." If only they embraced its true meaning ("happening one time in 10") then you could loosen your tongue and divulge the numbers of your inner heart. Until then, you remain a merchant of half-truths.

And now, back to you.

Philosopher

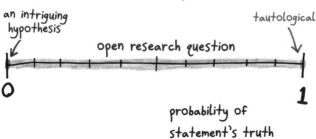

Hello! You are a philosopher. You read strange books, write even stranger ones, and enjoy walking into a bar alongside a priest and a rabbi.

Also, you do not let probability intimidate you.

Leave "probable" to the empiricists, you say. You seek ideas that are *im*probable. You ask questions no one else will ask and think thoughts no one else will think—mostly because they are pedantic, technical, and probably false. But this is to your credit! At your best, you birth new fields. Psychology, from Aristotle to William James, has its ancestry in philosophy. Even at your worst, your inquiries provoke and refresh us as no one else can.

Mission: Impossible Agent

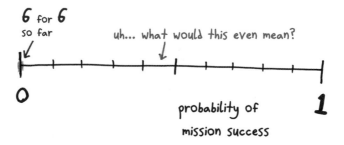

Hello! You are a *Mission: Impossible* agent. You dangle from ceilings into locked vaults, cling via suction cup to skyscrapers, and unmask your true identity to the double-crossers you've just triple-crossed.

Also, you do not seem to get what "impossible" means.

"Impossible" does not mean "as routine as a quarterly earnings report." Nor does it mean "very rare" or "rather difficult" or "whew, it's a good thing that knife blade halted a millimeter from Tom Cruise's eye." It means "not possible." And yet it keeps happening. Your film's title is as dishonest as its theme song is catchy.

In this, you are not alone. All fiction bends plausibility. One of my favorite shows is *Friday Night Lights*, a humble exploration of life in a small Texas town, depicting the human-scale struggles of ordinary people. Yet even this paragon of plausibility concludes each football game with once-in-a-lifetime dramatics: a 90-yard touchdown pass, a goal-line fumble, or a would-be field goal bouncing off the crossbar. It's enough to make me wonder: are we projecting our improbable fantasies onto the TV screen, or is the TV screen planting those fantasies in us to begin with?

Millennium Falcon Captain

|—— ~~NEVER TELL ME THE ODDS~~ ——|
0   **1**

probability of
successfully navigating
an asteroid field

Hello! You are the captain of the *Millennium Falcon*. You are a ruffian, a rascal, and a gold-hearted rogue. Your life partner is an 8-foot space dog wearing a bandolier.

Also, you reject probability altogether.

You're not a man of sober reflection and strategic consideration. You're a contraband smuggler, an empire toppler. You're a quick-draw Greedo-killing daredevil for whom doubt is fatal and hesitation is worse. There are no probabilists in foxholes, and you have spent your life in foxholes. For you, laborious probabilistic calculations would be as cumbersome as a neurotic golden robot who keeps saying, "Oh my!" and "If I may say, sir . . . "

I like to think that there's a bit of you in each of us. Probability serves us well in times of cool and careful assessment, but sometimes we need a confidence that's unwarranted by the hard quantitative facts. In moments of instinct and action, the soul leashed to plausibility may shrink from a necessary leap. At such times, we need to forget the numbers and just fly.

Chapter 14

WEIRD INSURANCE

COVERAGE SCHEMES FROM THE DEEPLY STRANGE TO THE STRANGELY DEEP

Despite my best efforts and worst drawings, I have become a grown-up. I drink unsweetened coffee. I wear unironic ties. Children call me "mister." Most discouraging of all, I spend a nonzero percentage of life thinking about insurance: health, dental, home, auto. Before long, I'll be insuring my ties against coffee stains, all without a shred of irony.

Beneath the suffocating weight of adulthood, there's something funny going on here. Insurance companies don't really sell products. They're vendors of psychology, offering peace of mind. They let short-run folks like me purchase a bit of stability from the long run of eternity.

So, for this chapter, let us spit our coffee back in the face of adulthood and reach for the chocolate milk instead. Let us gaze upon insurance with fresh eyes and explore the mathematics of this quirky industry, not through its traditional products, but through a series of peculiar and unconventional plans. Let us investigate the promise of mutual gain, the danger of exploitation, and the slow data-driven tug-of-war between the insurer and the insured—all while talking about college football and aliens.

Even in the darkest depths of grown-up-ness, it's worth remembering that insurance is a strange and silly thing. Kind of like adulthood itself.

SHIPWRECK INSURANCE

Ever since there have been ships, there have been shipwrecks. In China five millennia ago, merchants traded goods downriver in boats. If all went per usual, then the cargo would arrive, no problem. But if rapids and rocks intervened, you'd find your shipment rerouted to the riverbed, which can rarely be trusted to pay up.

Fortune dealt in two cards: moderate success and utter disaster. The former gave you reason to play the game, but the latter could ruin you.

What was an ancient Chinese trader to do?

Simple: redistribute the danger. Instead of loading one's own boat with one's own goods, merchants would scatter their wares across each other's ships so that each vessel carried a mixed cargo. By coming together, they turned the small likelihood of a high cost into the high likelihood of a small one.

This kind of risk redistribution predates history. You see it in any tight-knit community where people sacrifice to help their neighbors, knowing they might one day draw on the same communal kindness. A minor guaranteed cost helps eliminate an existential risk.

Today, "tight-knit community" has given way to "for-profit insurance company." Old-school social mechanisms of enforcement (e.g., not helping the stingy people) have yielded to precise formula-driven ones (e.g., setting rates based on risk factors). History has distilled a social arrangement into a mathematical one.

100 Boxes, 1 Boat

so risky!

Cargo lost	Likelihood
None	99%
All	1%

100 Boxes, 100 Boats

Cargo lost	Likelihood
None	36.6%
1/100	37.0%
2/100	18.5%
3/100	6.1%
4/100	1.5%
5/100	0.3%

Probably lose something. Won't lose everything.

Still, in principle, insurance today works much as it did 5000 years ago. I find fellow souls facing the same mortal risks. We exchange pieces of our livelihood. And we float downriver knowing that now, no shipwreck can sink us.

INSURANCE TO MAKE SURE THE KING HAS YOUR BACK

Quick time-travel jaunt. Set your watch to "Iran, 400 BCE; the New Year holiday of Norouz."

We arrive at the palace, where the monarch, flanked by bureaucrats and notaries, greets a long line of visitors. Each presents a gift, which circulates swiftly among the king's accountants so it can be assessed and documented. We are witnessing that most joyous of celebrations: insurance enrollment.

If you donate an item worth more than 10,000 darics, your name will land on a special honorary ledger. In your time of need, you shall receive 20,000 darics. Smaller gifts don't merit that guarantee but may still receive generous payouts. Take the humble fellow who donated a single apple and later received in return an apple full of gold coins.

Deluxe fruity piñatas aside, the scene invites a more cynical way to conceptualize insurance. It's the rich selling a bit of resilience to the poor.

When bad things happen to rich people, they cope. If you're a king, a billionaire, or a grandchild of Sam Walton, then you pay the hospital bills, buy a new car, and move on with life. But when bad things happen to poor people, they struggle. They might forgo medical treatment, take on debt, or lose a job because they can't afford transportation to work. The rich rebound after misfortune; the poor spiral.

Thus, in a step of topsy-turvy logic, insurance is born. The poor, unable to afford risk, pay the rich to take it off their hands.

"OH NO, MY EMPLOYEES ALL WON THE LOTTERY" INSURANCE

It's a common scene: workers pooling their money to buy lottery tickets, hoping to quit their jobs en masse. They are comrades unified by the dream of no longer being unified. But suppose they win. What happens to their poor boss, awaking one morning to the abrupt resignations of an entire department, like some kind of unnatural disaster?

Fear not, forsaken boss. There's a cure for what ails you.

In the 1990s, over a thousand British businesses invested in the National Lottery Contingency Scheme. If your staff won the jackpot, insurance payouts would help to cover recruitment, retraining, revenue losses, and temporary labor.

(I reckon the government, which runs these lotteries, should sell the insurance plans, too. It'd be like Godzilla Insurance Co. offering fire-breath protection.)

As you'd guess, the expected value favors the insurer. What might surprise you is the margin. The odds of a single ticket winning the jackpot are 14 million to 1, and yet the payout ratios max out at 1000 to 1. Smaller businesses fare even worse. With two employees, the expected value remains unfavorable even if each staffer buys 10,000 tickets.

to equalize expected value

Payout -to-Premium Ratio	Employees Covered	Tickets per Employee
1000 to 1	100	140
500 to 1	2	14,000

best deal
(£300 premium)

worst deal
(£50 premium)

Of course, insurers can't just charge the expected value. They've got to consider their own costs, not to mention the risk involved. An insurer that sets its margins too narrow will go bankrupt the first time a slightly unusual number of clients demand payouts.

As for the fearful bosses, is there a better option? Well, if you can't insure 'em, join 'em. Tossing a little money into the pool is an insurance plan in itself: a small investment that will pay out if (and only if) they win.

MULTIPLE-BIRTH INSURANCE

Wait.. but if I'm holding our newborn, _and_ you're holding our newborn...

OH MY WORD

THERE'S TWO OF THEM

GET HELP

I view "one" as a formidable number of babies to care for. "Two" sounds like quartering a child army. "Three" is unimaginable—a childbirth googol. So it's dizzying to learn that 1 in 67 pregnancies results in multiple births. In the UK, expectant parents who share my trepidation can purchase "multiple-birth insurance."

You pop out twins, and the insurance company pays out £5000.

One analyst and personal finance adviser, David Kuo, scoffs at this plan, saying: "You're better off putting money on the horses." I see his point: The expected value here is pretty lousy. The lowest premium is £210, for a mother with an age below 25 and no family track record of twins. If a hundred such mothers buy the insurance, then the company will collect £21,000, and (most likely) pay out just one sum of £5,000. That's a comfy profit margin, which is the polite way to say "their customers are paying through the nose."

Parents with a greater likelihood of twins fare little better. One 34-year-old mother who was a twin herself paid £700 for coverage. That's a payout ratio of just 7 to 1.

If you've had fertility treatments, or waited until after the 11-week ultrasound, then you're not eligible at all.

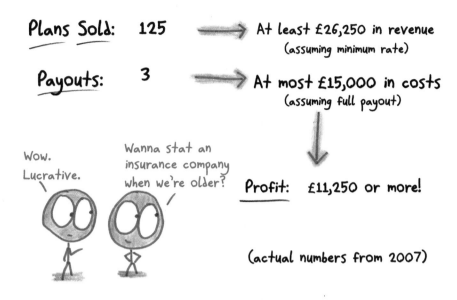

Plans Sold: 125 → At least £26,250 in revenue
(assuming minimum rate)

Payouts: 3 → At most £15,000 in costs
(assuming full payout)

Wow. Lucrative.

Wanna stat an insurance company when we're older?

Profit: £11,250 or more!

(actual numbers from 2007)

I find it helpful to distinguish between two types of insurance: *financial* and *psychological*.

Most common insurance products—health, life, auto, etc.—hedge against financial ruin. But psychological insurance hedges against more modest risks. Take travel insurance. You can probably afford the plane tickets (heck, you've already paid for them), but no one feels good canceling a vacation. By offsetting the financial cost, insurance can ease the emotional one.

This isn't a self-help finance book (*Math with Bad Investment Advice*). Still, I shall offer one piece of counsel: Be wary of psychological insurance.

Imagine if the multiple-birth insurer offered a raft of other plans. Would you pay £200 to insure against postpartum depression? Or the higher cost of raising an autistic child? Or a child born with Down syndrome? Or a kid with a chronic illness? Each risk feels worth protecting against. But buy enough plans at £200 each, and eventually, the cumulative cost of the premiums will swamp any payout. If you can afford the insurance, then you can afford the outcome it's insuring against.

A more frivolous example: For $3 each, I'll offer to insure any T-shirt worth $20 or less. Eager to protect your wardrobe, you insure 15 beloved favorites. Then, when your dogs and/or twins chew one up, I'll ship a replacement. However, if this happens once per year, you're paying $45 for a service worth no more than $20. Why not just keep the money and replace your own T-shirts?

Cost: $45

Payout: $20

It's the basic problem of all psychological insurance. Why pay good money to offload a risk that you could handle yourself?

ALIEN-ABDUCTION INSURANCE

In 1987, a fellow in Florida named Mike St. Lawrence rolled out a $19.95 alien-abduction policy. It provided $10 million in coverage, paid out at a rate of $1 per year (until completion or death, whichever comes first). It also entitled abductees to "psychiatric care" and "sarcasm coverage" ("limited to immediate family members"). The slogan: "Beam me up—I'm covered." Applicants were rejected if they answered yes to the question "Do you take this seriously?" or "Were your parents related before they married?" It was hard to miss the winking.

Then a British company began selling abduction coverage in earnest. Soon, they had sold 37,000 plans at £100 apiece—and (surprise!) not made a single payout. That's nearly £4 million in pure profit. Said one managing partner: "I've never been afraid of parting the feeble-minded from their cash."

The silly story points toward a real danger. Psychological insurance can quickly turn predatory. When we buy protection against fear itself, we invite companies to make us afraid.

FAILED-EXAM INSURANCE

During your mathematics PhD at UC Berkeley, you take a high-stakes qualifying exam known as "the Qual." This is a monster you face alone: three hours of oral questioning from a trio of expert professors. Though preparation is exhausting, the passing rate is high—at least 90%—because it isn't scheduled until you seem battle-ready. The expert professors desire a retake no more than you do.

Qual failure, then, fits the following criteria: (1) it's rare; (2) it hurts; (3) many individuals face it; and (4) it strikes more or less at random. That would seem to make it ripe for insurance.

When Should I Sell Insurance?

A Checklist for Entrepeneurs

— Hmm... what about "disappointing TV show finale" insurance?

1. The risk is rare.

Not as rare as I'd like.

2. The risk causes suffering.

Undoubtedly.

3. The risk threatens many individuals.

Millions!

4. The risk strikes individuals more or less at random.

No... it targets fans of overdeveloped mythologies.

Some PhD students have hashed out the idea. Each participant pays $5 into the pool. If anyone fails their exam, then they collect the whole bundle. If everyone passes, the money goes toward a round of celebratory beers.

I suppose every financial instrument designed by grad students must culminate, one way or another, in beer. It's psychological insurance of the best kind.

HOLE-IN-ONE PRIZE INSURANCE

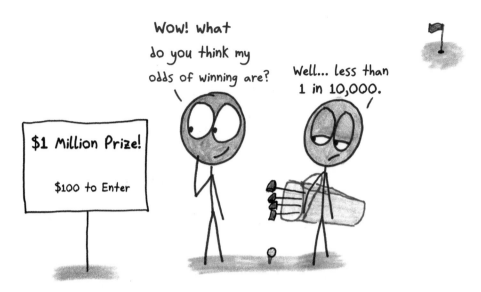

There's nothing more American than a gaudy competition. The $10,000 half-court shot. The $30,000 dice roll. The $1 million hole in one. When rolling out such promotional gimmicks, there's just a teeny-tiny risk.

Someone might win.

As it turns out, this market is a dream for insurance companies. As merchants of probability, their solvency depends on accurate computation: 50-to-1 payouts on a 100-to-1 event will keep you in the black; 100-to-1 payouts on a 50-to-1 event will bankrupt you. The actuarial intricacies of home, life, and health insurance make it easy for an insurer to miscalculate.

But prize giveaways? No problem!

Take the hole in one. Amateur golfers succeed roughly once in every 12,500 attempts. So, for a $10,000 prize, the average golfer costs $0.80. Charge $2.81 per golfer (one company's advertised premium), and you'll turn a tidy profit. The same company insured a $1 million hole-in-one prize for just $300. It's a fine deal on both ends: the charity offloads risk, and the expected-value cost to the insurer is just $80.

The company also insured an NBA fan's $16,000 half-court shot for $800. Unless the fan's odds of success exceed 1 in 50—about what NBA players manage in game situations—it's a square deal. Or consider insuring Harley-Davidson against a $30,000 prize given out to a customer who rolls six lettered dice to spell out *H-A-R-L-E-Y*. The expected value is just $0.64 per contestant. The company charged $1.50. Actuarial math has never been easier.

Payout Probability: 12,500 to 1

Payout Ratio: 3500 to 1

Payout Probability: 200 to 1

Payout Ratio: 50 to 1

Payout Probability: 46,656 to 1

Payout Ratio: 20,000 to 1

That's not to call this type of insurance riskless. In 2007, the Boston-area retailer Jordan's Furniture ran a glitzy promotion: buy any sofa, table, bed, or mattress in April or May, and if the Red Sox go on to win the World Series in October, then the price will be refunded. The sales approached 30,000 items—a total of perhaps $20 million.

Then the Red Sox won. The Sox-loving folks at Jordan's were thrilled; they'd bought insurance.

The folks at the insurance company? Less delighted.

"CHANGE OF HEART"
WEDDING INSURANCE

Please join me now for a thought experiment. A stranger hands you a coin. You flip it and hide the result. What's the probability of heads?

- *A random passerby*: "I don't know . . . 50%?"
- *The deviant who gave you the coin and knows that it's biased toward heads*: "70%!"
- *You, having snuck a look at the outcome*: "100%."

No one here is wrong. Uncertainty is a measure of knowledge, and each person has given a rational, well-calibrated response to the available information. The fact is, every probability ought to come with a footnote: "contingent on what I know now."

This dynamic can become a nightmare for insurance companies. What if they're the random passerby, whereas the person buying the insurance has snuck a peek at the coin?

The company Wedsure insures against all kinds of wedding fiascoes: damaged bridal gowns, stolen gifts, no-show photographers, mass food poisoning. But the most sensational coverage option they offer is also the most problematic: the "Change of Heart" clause.

If anyone knows of a reason why these two should not wed, speak now, or forever forsake your insurance claim.

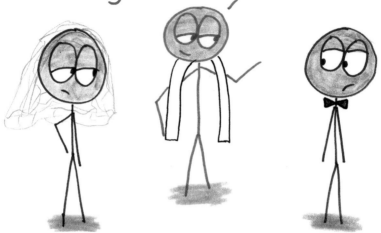

In 1998, a journalist asked Wedsure owner Rob Nuccio if the company would consider reimbursing for weddings canceled because of cold feet. He scoffed. "That would too easily set us up for fraud and collusion," he said. "If you're asking the question, don't get married." By 2007, he had changed his mind and begun offering to reimburse third-party financiers (but not the couples themselves) for cancellations with at least 120 days' notice. But according to Nuccio, a problem arose: "We were getting claims from mothers of the bride who . . . knew in advance that there was a problem." Obviously, each woman had mapped her daughter's heart better than the insurance company could. They had peeked at the coin. Nuccio widened the time window to 180 days, then 270, and eventually 365.

It's a quintessential problem for insurers. The person seeking coverage often knows details that the provider cannot. The insurer has five possible solutions:

1. Set your profit margin even higher than normal.
2. Employ a detailed application process to equalize the knowledge level.

3. Offer a cheap low-coverage option, and a pricey high-coverage option. The high-risk people will naturally be drawn to the latter.
4. Become so expert in the risk you're insuring that you know more than your customers.
5. Stop insuring that risk.

INSURANCE-COMPANY INSURANCE

What do you mean, you won't pay? We have insurance!

Hey, we never guessed a hurricane would strike you and all your neighbors on the exact same day!

Knowledge gaps scare insurers, but they don't jolt them awake in a cold sweat. Their true wide-eyed nightmare is a single whispered word: *dependence*.

Think back to the Chinese traders, who scattered their merchandise across a hundred boats. What if shipwrecks don't come one by one? What if on 99% of days, no boats sink, and on 1% of days, a storm sinks them all? Then insurance is useless. We're not facing individual risks that can be mitigated by redistribution. We're all holding the same awful ticket to the same terrible lottery. Trading pieces of fate does nothing. When we go down, we go down together.

That's dependence. For insurers, it's doom. Hurricane Katrina, for example, exacted its $41 billion in damage all at once, far exceeding the year's $2 billion or so in premiums. In such cases, the insurance company faces the same risk as the people it's insuring. The eggs remain all in one basket.

The solution is to scale up: insure the insurance company. This is called "reinsurance." Insurers from different regions trade assets, in effect swapping customers. This lets a company in one location assemble a risk portfolio that's spread far and wide.

When it's not enough to scatter your livelihood across a hundred boats, then scatter it across a hundred rivers.

INSURANCE FOR COLLEGE FOOTBALL PLAYERS IN CASE THEY GET WORSE

Time for some fantasy role-play: you are a college football star. Grown men paint their bare chests in your team's colors. You've got skills worth millions. But you can't collect on the payday while you're in college—and devastating injury can strike at any time. It's like your body is a jackpot lottery ticket you're forbidden to cash in until you've run it through the washing machine a few times. Better hope those winning numbers don't fade.

What's your move, kid?

You can insure against career-ending injury, but that's not enough. What if an injury *hampers* your career without *ending* it? What if it drops you from the first round of the draft (for a contract worth $7 million) to the sixth round (for a contract worth $1 million)? More than 80% of your payday has gone up in smoke, and that pricey insurance plan (costing $10,000 per $1 million of coverage) does nothing.

This is why top-tier players have begun purchasing an additional plan: "loss of value" insurance. It's not cheap; it costs an additional $4000 in premiums per $1 million of coverage. But for high-value prospects, it's worth it. Players like cornerback Ifo Ekpre-Olomu and tight end Jake Butt have already collected. More, no doubt, will follow.

HEALTH INSURANCE

I've saved for last the weirdest risk-reducing product of all: health insurance.

What's so strange about it? First, the complexity. Deductibles, coverage limits, preexisting conditions, co-pays, variable premiums . . . It's not quite brain surgery, but it *is* the financing of brain surgery, which is perhaps just as daunting. And second, there's the vexing question of whether, as our medical knowledge and predictive powers grow, health insurance can continue to function at all.

Here's a simple model of healthcare. We each flip 10 coins. If you flip all heads, then you contract the dreaded "ten-headed disease," which will require $500,000 in medical bills.

Each of us faces a roughly 1 in 1000 risk of disaster. But if we all chip in $800 for insurance, then it works out. Per thousand people, the insurance company collects in $800,000 and pays out just $500,000. They turn a profit; we hedge against financial ruin. Everybody goes home happy.

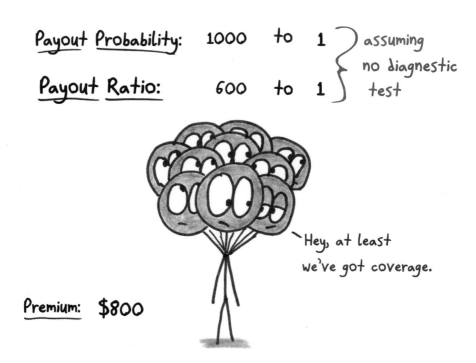

Payout Probability: 1000 to 1 ⎞ assuming

Payout Ratio: 600 to 1 ⎠ no diagnostic test

Hey, at least we've got coverage.

Premium: $800

But what if we know more? What if, before deciding whether to purchase the plan, we can peek at the first five coins?

Now, from every 1000 people, roughly 970 of us glimpse a tail and sigh with relief. We're safe; no insurance necessary. But the remaining 30 people grow nervous. One of them probably has the dreaded disease. Between them, those 30 can expect a collective $500,000 in expenses. Even if they redistribute the costs equally, each remains on the hook for thousands and thousands of dollars. Insurance no longer provides peace of mind; it drains it.

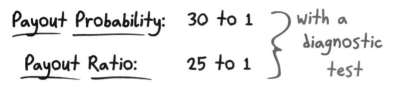

When we looked at the coins, we reduced our uncertainty, and without uncertainty, insurance collapses. If we know in advance who will suffer— whose boat will sink, whose employees will win the lottery, who will suffer that NFL-dream-crushing injury—then insurance becomes impossible. You can only pool risk among people who share it.

In our medical system, this problem is unfolding year by year. Genetic testing and improved statistics threaten the basic logic of insurance.

I see no easy solution. If you individualize the rates, then some folks pay pennies while others face premiums almost as steep as the medical bills themselves. But charging everyone the same amount turns a project of mutual benefit, with each of us hedging against risk, into a project of collective charity, with some folks subsidizing others. That's a harder sell. It's one reason why American healthcare remains so contentious.

As a teacher, I'm inclined to think of all knowledge as a gift. But insurance complicates that story. Our ignorance of fate can force us to collaborate against it. We have built a democracy from our uncertainty—and new knowledge threatens that balance, as sure as flood or fire.

Chapter 15

HOW TO BREAK THE ECONOMY WITH A SINGLE PAIR OF DICE

1. THE HAUNTED CAREER FAIR

In September 2008, I began my final year of the great free-pizza scavenger hunt called "college." Knowing, if not quite believing, that after college only wage earners could acquire pizza, I decided to attend the annual career fair. Booths of employers would fill the gymnasium to offer free giveaways and (secondarily) job applications.

But when I arrived, I found a half-empty gymnasium, a ghost town. In abrupt unison, the investment banks had all decided that maybe now wasn't such a good time to hire.

We knew why. In the prior month, the global financial system had frozen, tossed up a blue screen of death, and was now refusing to reboot. Wall Street had devolved into the final scene of a Shakespearean tragedy: century-old institutions lay heaped in the dirt, swords sticking out of them, wheezing death monologues. Journalists batted around terms like "worst," "recession," and "since the Great Depression," often consecutively. Even the pizza crusts tasted of anxiety.

In this chapter, we come to our final lesson on probability, perhaps the hardest. Many would-be probabilists fall for the seductive idea of *independence*, choosing to imagine our world as an aggregation of isolated events. But if probability is to confront the world's uncertainty, it must face the world's interconnectedness: its narrative threads, its causal chains.

A simple illustration: What's the difference between rolling two dice and doubling a single die?

Well, in each case, the sums range from a minimum of 2 (snake eyes) to a maximum of 12 (double sixes).

With two independent dice, extreme values can unfold only a few ways. (For example, only two combinations will yield a 3). Middle values can unfold in many ways; for example, six different combinations will yield a 7. Thus, middle values are more likely.

What about rolling just one die and doubling it? Now the "second" roll depends entirely on the first; we've got a single event disguised as two. And so the extreme outcomes are just as likely as the middle ones.

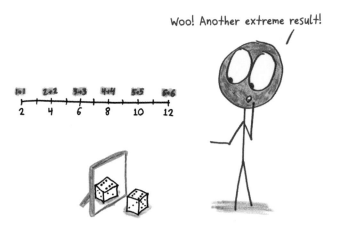

The difference is stark. Independence irons out extremes; dependence amplifies them.

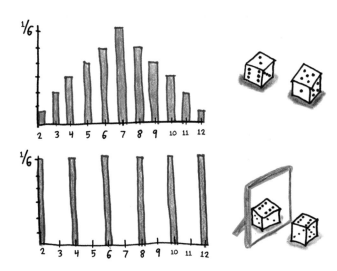

We can push it further. Let's move from two dice to a million. Now, the outcomes range from a minimum of 1 million (all ones) to a maximum of 6 million (all sixes).

What if each die acts alone, independent of the other 999,999? Then we find ourselves in the long-run world of stable tendencies, where the exciting 6's and the disappointing 1's arrive in equal proportions. The overwhelming likelihood is that our total for all the dice will land smack in the middle, far from both extremes. There's a 99.9999995% probability we'll land between 3.49 million and 3.51 million. Getting the minimum of 1 million is all but impossible, its probability less than 1 in a googol googol googol googol. (I could write the word "googol" another 7000 times, but you get the idea.)

But what if we're not rolling separate dice? What if we're rolling a single die and multiplying it by a million, reflecting it through a hall of mirrors? Well, then we remain in the chaotic world of random chance. The later "rolls" depend wholly on the first, providing no balance. Getting the minimum is no longer an absurd when-pigs-fly proposition. Its probability is 1 in 6.

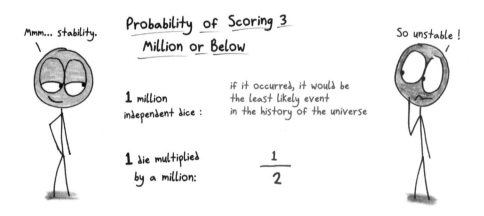

Mmm... stability.

So unstable!

Probability of Scoring 3 Million or Below

1 million independent dice :   *if it occurred, it would be the least likely event in the history of the universe*

1 die multiplied by a million: $\dfrac{1}{2}$

Insurance, portfolio diversification, and egg-in-basket placement all rely on the same fundamental principle: Overcome risk by combining events. One stock is a gamble; combine many of them into an index, and you've got an investment.

But all of this depends on independence. It's pointless to scatter your eggs across several baskets, then fasten the baskets together and load them into the bed of the same pickup truck. A world of dependence is a world of feedback loops and domino cascades, a world of extremes. It's a world where career fairs look like either festivals or funerals, with little in between, a world where banks all thrive until they abruptly go down together.

2. EVERYTHING HAS A PRICE

Pop quiz! What is the basic activity of Wall Street banks?

A. Powering the world economy via the intelligent allocation of capital
B. Buying Italian suits with blood money snatched from the pockets of the working class
C. Pricing things

If you answered A, then you work for Wall Street. (Hey, nice suit! Is that Italian?) If you answered B, then I'm honored that you're reading my book, Senator Sanders. And if you answered C, then you're already familiar with a key theme of this chapter: that the basic function of the financial sector is to decide what things are worth. Stocks, bonds, futures, Rainbow Shout® contracts, standard Parisian barrier options, single-use monocular default swaps . . . whether you're buying, selling, or Googling to see which of those I made up, you want to know what these things are worth. Your livelihood depends on it.

The problem, of course, is that pricing ain't easy.

It's a CDS for the second tranche of a CDO, but remember that TARP could make your Great Barrier Reef options more valuable than Algerian lookbacks from Gringotts. So what'll you pay?

Um... $5?

Take bonds. These are pieces of debt, promises that you'll get paid back. Consider somebody who borrowed money to buy a house and promises to pay back $100,000 in five years.

What's this IOU worth to you?

Well, we begin with Pricing Challenge #1: putting a value on *time*. In the carpe diem logic of finance, a dollar today is worth more than a dollar tomorrow. First, there's inflation (which gradually lowers the value of a dollar), and second, there's opportunity cost (i.e., a wisely invested dollar today will be worth more by next year). As a crude estimate, we'll say a dollar today is worth $1.07 next year. Carry that forward, multiplying year after year, and you'll find that $1 today is equivalent to $1.40 in five years.

Years from Now	Value of $1 Today
0	$1.00
1	$1.07
2	$1.14
3	$1.23
4	$1.31
5	$1.40

Huh. Better get my dollars now, so they can grow.

Receiving $100,000 in five years isn't as glamorous as it sounds. It's like getting paid just $71,000 today.

Is that figure the bond's true price? Can we call it a day and go wash that icky Wall Street feeling off our hands? Alas, we're just getting started. We must also put a price on *risk*: Who owes us the debt, and can we count on them? If it's a two-income household with perfect credit history and sparkling teeth, then the odds are with us. But if our debtor is someone shiftier—say, a pizza-addicted recent college grad with a penchant for bad drawings—then there's a real chance our bond will amount to nothing.

How do we adjust the price?

Simple: expected value. If there's a 90% chance of getting paid back, then the bond is worth 90% as much.

Likelihood of Being Repaid	Value of the $100,000 Bond
100%	$71,000
90%	$64,000
75%	$53,000
50%	$36,000
25%	$18,000
10%	$7,000

Makes sense. Half the probability, half the price.

We're still not done. Default isn't binary, with debtors paying back all or nothing. In reality, courts and lawyers will intervene to hammer out a deal in which creditors receive some fraction of what they're owed, ranging from pennies per dollar to almost the whole sum. Our bond, then, is like a bizarro high-stakes lottery ticket. How do you put a single price on such variety?

Again: expected value. We make educated guesses about how much we might get paid back, and then we imagine buying millions and millions of such bonds. Instead of straining to divine the unknowable price of this specific bond, we compute the long-run average of all such bonds.

Percentage Repaid	Value of Bond	Probability
100%	$71,000	0.5
80%	$57,000	0.1
60%	$43,000	0.1
40%	$29,000	0.1
20%	$14,000	0.1
0%	Nothing	0.1

Average : $50,000

And there you have it. A $50,000 bond.

On Wall Street, pricing is like breathing: routine, constant, and essential for survival. But for decades, just about the only goods that banks felt comfortable evaluating were stocks (pieces of companies) and bonds (pieces of debt). That excluded "derivatives," which are neither stocks nor bonds but their mutant descendants. They lived at the fringes of the finance industry, like a shady alleyway casino flanking the respectable bank.

With the 1970s came a sea change: "quantitative analysis." With the power of mathematical modeling, "quants" figured out how to price derivatives—even those of Dr. Seuss–level weirdness. Among the most complicated of all were CDOs: collateralized debt obligations.

Although they came in many varieties, a common recipe went something like this:

1. Gather thousands of mortgages (like the one we priced earlier) into a single bundle.
2. Slice the bundle into layers (called "tranches"), ranging from "low-risk" to "high-risk."
3. When interest payments come pouring in, the owner of the low-risk tranche gets paid first, and the owner of the high-risk tranche gets paid last.

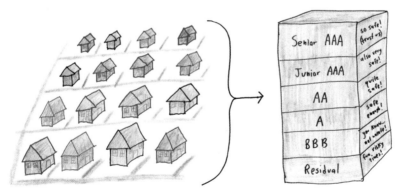

CDOs offered an à la carte menu of risks and rewards, a tranche to match any taste. Willing to pay extra for a safe bet? Consider this delicious upper tranche. Looking for a cheaper higher-risk option? Then I've got a spicy lower tranche you'll just love. You prefer something in between? Well, just let our chefs know, and I'm sure they can prepare it to order.

Investors were smacking their lips and banging on the table for more . . . right up until September 2008, when the bill came.

3. THE PROBLEM OF THE HOUSE

Flash back to 1936, when surrealist painter René Magritte drew a series called "Le Problème de la Maison." The sketches show houses in peculiar spots: nestled in tree branches, hiding in cliffside caves, piled at the bottom of a giant ditch. In my favorite, a house stands upright in an empty plain, looking ordinary except for its two neighbors: a pair of gargantuan dice.

(do not be deceived by my artistic prowess;
this is a near-perfect facsimile, not the original)

Who knows what Magritte meant by it? This is the man who once painted a bird grabbing a lady's shoe and titled it *God Is No Saint*. But I think he's challenging our idea of the home as an emblem of security. Instead—and I don't mean to spook you—the house is something chancy and precarious, an existential risk. Your house will likely be the largest investment of your lifetime, costing several times your annual salary, incurring a debt that will last a generation. The house is an icon not of stability but of probability.

Seven decades after Magritte's visual pun, Wall Street faced its own little *problème de la maison*: pricing CDOs. The problem was to identify the relationship between the various mortgages. Last I checked, you and I are separate people. So if I default, perhaps that has no bearing on you. On the other hand, we share the same economy. We can escape a big recession no more than we can escape summer pop songs. So if I default, perhaps that suggests you're in danger too. In Wall Street terms, the question was whether defaults pose an *idiosyncratic* risk or a *systematic* one.

Are houses like thousands of separate dice, each roll independent of the others? Or are they like a single die, multiplied thousands of times by a hall of mirrors?

Imagine a CDO (tiny by real-world standards) built from 1000 of the mortgages we explored above. Since we priced one at $50,000, the whole bundle should be worth $50 million.

Now, if the mortgages are independent, Wall Street can sleep easy. Sure, our investment might come in $1 million below what we expected, but it would take absurd misfortune to lose $2 million, and a loss of $5 million is unthinkable, its probability less than 1 in 1 billion. Independence stabilizes, all but eliminating the chance of catastrophic loss.

If totally independent...

Investment Value	Probability of That Value or Less
$49 million	13%
$48 million	1%
$47 million	0.02%
$46 million	0.00008%
$45 million	0.00000008%

Ah, sweet stability.

Meanwhile, if all the houses represent a single roll of the dice, then Wall Street shall suffer night sweats and scream dreams. Dangers unthinkable a moment ago are now very, very thinkable. In this deal, there's a horrifying 1-in-3 chance we lose nearly half of our investment, and a bone-chilling 1-in-10 chance that we lose everything.

If totally dependent...

Investment Value	Probability of That Value or Less
$43 million	40%
$28 million	30%
$14 million	20%
Nothing	10%

This could go bad. Very bad.

Of course, neither vision quite matches reality. We are neither hive-minded drones living in perfect sync, nor rugged individualists unaffected by our neighbor's weather. Instead, our lives are somewhere in between, our futures delicately intertwined. It seems obvious that one default heightens the likelihood of another—but by how much, and under what conditions? These are the subtlest challenges that probabilistic models can face.

If partially dependent...

Investment Value	Probability
? ? ?	? ? ?
? ? ?	? ? ?
? ? ?	? ? ?
? ? ?	? ? ?

Wall Street's solutions included the notorious "Gaussian copula." This formula originated in life insurance, where it helped to adjust the probability of one spouse's survival after the other spouse's death. Replace "spouse" with "house," and "death" with "default," and you've got a model for calculating the dependencies between mortgages.

The formula captures the relationship between two mortgages with a single number: a correlation coefficient between -1 and 1.

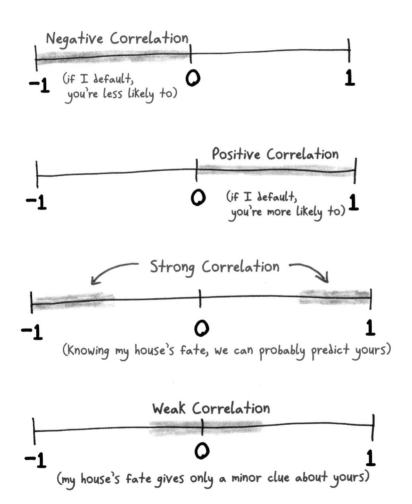

To its credit, the copula is a nifty piece of mathematics, simple and elegant. But the global economy doesn't really go in for simplicity, and in hindsight, it's easy to see what the copula (and similar approaches) missed.

First, the data. Wall Street's computers feasted on spreadsheets full of housing prices. But most of the figures came, in city after city, from the same recent time period, during which (it just so happened) home prices had enjoyed a steady rise. The models calibrated their confidence as if they had studied the entire history of US housing. In fact, they had witnessed a single roll of the dice, multiplied by a million.

Second, the model was built for couples (hence "copula"). But houses don't come in pairs. They exist, all together, as a loose nationwide market. A single change can affect every mortgage in the country, all at once—which is what happened when the soaring market fell to earth. It's silly to worry about one domino knocking over its neighbor, when there's a single giant domino looming above them all.

You better not knock me over!

Hey, don't knock me over!

Finally, if you know your statistical vocabulary, alarm bells start ringing at the term "Gaussian." In mathematics, the word arises whenever you add together lots of independent events. But that's the whole problem: these events weren't independent.

In all these ways, Wall Street overlooked the risk that Magritte saw, leading to a calamity as surreal as anything the painter ever envisioned. And still, we haven't discussed the worst of it. The mispriced CDOs amounted to a few trillion dollars—enough to damage the economy but not enough to explain the body blow it suffered in September 2008. What left the economy collapsed on the mat, dizzy and spitting teeth, wondering in bewilderment where that sucker punch had come from? It was a probabilistic failure on an even grander scale.

4. THE $60 TRILLION DOUBLE-OR-NOTHING

As you'll know if you read the previous chapter—a laugh-out-loud joyride for the whole family—insurance makes sense. I can't afford to lose my house (where would the pizza deliveries come to?), so I'm willing to pay a little premium each month, in exchange for a big payout should my home ever burn down. I eliminate an existential risk; the insurance company turns a profit. Everybody wins.

But here's a weird notion. What if *you* insure my house?

You can mirror the bargain I made. Small regular payments, leading to a big windfall should disaster strike. This duplicates the financial structure of insurance, but without the purpose. It's just a wager, a win-lose zero-sum game. If my house burns down, you win; if my house stays safe, the insurance company wins. Even weirder: what if *thousands* of people hop on board, each buying insurance on my house, standing to win if it burns down?

Nice house you got there.

Shame if something happened to it.

Hey, kid, want some matches?

I may fret as I begin to receive anonymous gifts, such as cheap fire-crackers and working grenades. But I'm not the only one losing sleep in this scenario; the insurance company is even more freaked-out than I am. If my house burns, they're left holding fistfuls of ash, on the hook for thousands of enormous payouts.

That's why no insurance company would agree to the situation. The 95% chance of easy profit cannot offset the 5% chance of utter ruin.

It's too bad no one pointed this out to Wall Street.

For Wall Street, doom came with a three-letter acronym: CDS. It stood for "credit default swap," which was basically an insurance policy on a CDO. You make modest, regular payments. As long as the CDO keeps paying interest, nothing happens. But if enough mortgages default, then the CDS kicks in with a big payout.

So far, so reasonable. But guess what Wall Street did next? They sold dozens of CDSs for each underlying CDO. It's like selling dozens of insurance policies on the same house, except with a lot more zeroes on the end. By the start of 2008, the amount of money on the line had reached $60 trillion. That is roughly the GDP of the planet Earth.

A quick recap. CDOs were meant to achieve the stability of a million dice; instead, they embodied the fragility of a single die. The CDSs doubled down until the gamble loomed large enough to imperil the entire world economy. It all prompts a natural question:

How could Wall Street have been so stupid?

You harvest the greatest minds from the fanciest universities, buy them million-dollar supercomputers, pay them cosmic salaries, work them 90 hours a week . . . and then walk in to find them screaming and jamming forks in the wall sockets?

I wish I could write off these "independence vs. dependence" errors as a one-time aberration, a peculiar circumstance of CDOs and CDSs. But if wishes were houses, the CDOs might have stood a chance. The grim fact is that this error runs to the core of financial markets.

5. ASHES, ASHES, WE ALL FALL DOWN

At the risk of being dubbed a neoliberal shill, I will state my belief that markets work pretty well. Heck, I'll go further: they work *really* well.

For example, this planet happens to harbor delicious fruit spheres called "apples." How should we allocate them? If farmers grow more than consumers want, we'll find piles of Red Delicious rotting in the streets. If consumers want more than farmers grow, we'll witness strangers clawing at one another to snag the last McIntosh during a shortage. Yet somehow, against all odds, we manage to grow just about the right number of apples.

The trick? Prices. Although we think of prices as determining our behavior ("too expensive, so I won't buy"), the reverse is just as true. Each of our individual choices exerts a tiny influence on the price. If enough people refuse to buy, the price will drop. If enough refuse to sell, the price will rise. The price functions as an aggregate of all our independent judgments and decisions.

And so, like other aggregates of independent events, prices tend to yield balanced, stable, reasonable outcomes. Aristotle called it "the wisdom of the crowd." Adam Smith called it "the invisible hand." I call it "the independent dice all over again, except this time, the dice are us."

In theory, what works for apples ought to work for CDOs. Some folks will overvalue them. Others will undervalue them. But in the end, a market full of independent investors will drive the price to a stable equilibrium.

There's just one problem: all too often, investors behave less like millions of independent dice, and more like a single die multiplied by millions.

Take the 1987 stock crash. On October 19, prices nose-dived, plummeting more than 20%. It came without warning: no market-shaking news, no high-profile bankruptcy, no "Oh jeez, I'm in over my head" speech from the Fed chair. Markets just collapsed. Only later did the postmortem identify a peculiar trigger: lots of Wall Street firms had relied on the same basic theory of portfolio management. Many even employed the same software. As markets slipped, they sold the same assets in unison, sending prices into a downward spiral.

Portfolio management's whole purpose is to bring safety through diversification. But if everyone diversifies in exactly the same way, then the resulting market ain't so diverse.

If investors exercise their own judgment, then day-to-day price changes should follow a bell-shaped normal distribution: up a little on some days, down a little on others, but almost never leaping too fast or far. Alas, reality disagrees. Market movements, with their occasional massive drops, follow more of a *power-law distribution*. That's the same mathematical model we use for earthquakes, terrorist attacks, and other massive disruptions to highly sensitive systems.

The market isn't random like lots of dice added together. It's random like an avalanche.

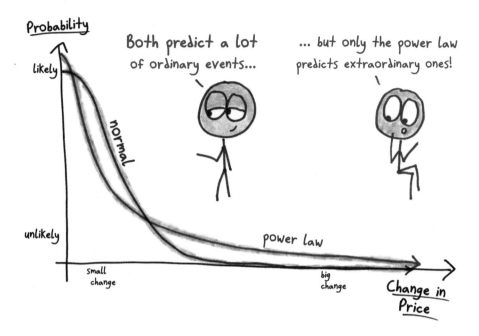

Leading up to the financial meltdown of 2008, many banks relied on the same limited set of models (such as the Gaussian copula). Rather than bringing fresh insight, they herded around a single strategy. Even ratings agencies—whose purpose and duty is to offer independent analysis—just parroted the banks' own claims. The referees became cheerleaders.

Why, back in September 2008, did I arrive to find a somber half-full gymnasium? Why, in other words, did the financial system melt down?

Well, it's complicated: Incompetence played its part, as it does in most failures. (Just watch me bake sometime.) So did misaligned incentives, blind optimism, naked greed, dizzying complexity, governmental dysfunc-tion, and interest rates. (Again, see my baking.) This brief chapter has told only a sliver of the story, focused around a specific theme: the dangerous assumption of independence, when dependence actually reigns.

Mortgages default together. CDSs pay out together. And actors in the marketplace herd together around similar pricing strategies.

You want to break the economy with a single pair of dice? Honestly, it's easy. Just convince yourself that you're rolling a million pairs, and then bet your fortune on the roll.

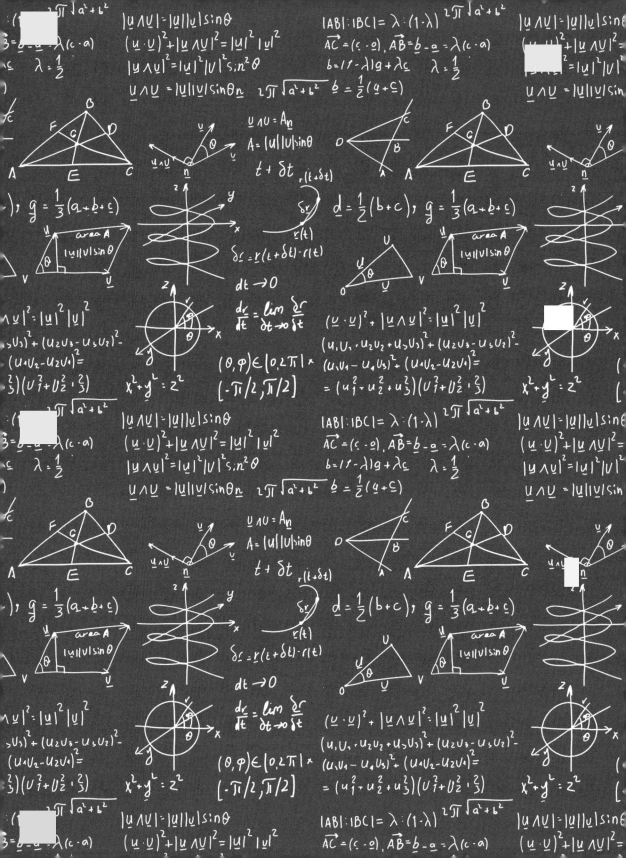

IV

STATISTICS

THE FINE ART OF
HONEST LYING

A survey of medical professionals asked them to contrast clinical evaluation with a more statistical approach. These are the words they used to describe each method:

CLINICAL METHOD IS...

Dynamic   Patterned   Real

Global   Organized   Living

Meaningful   Rich   Concrete

Holistic   Deep   Natural

Subtle   Genuine   True to life

Sympathetic Sensitive   Understanding

Configural   Sophisticated

Life is a web.
Existence is one.

STATISTICAL METHOD IS...

Mechanical	Arbitrary	Superficial
Atomistic	Dead	Rigid
Additive	Pedantic	Sterile
Cut and dried	Fractionated	Academic
Artificial	Trivial	Pseudoscientific
Incomplete	Forced	Blind
Unreal	Static	

Life is mere dust on the carpet of existence

On behalf of statisticians everywhere: ouch.

I admit that there is something reductive about the whole project of statistics, of taming the wild, unpredictable world into docile rows of numbers. That's why it's so important to approach all statistics with skepticism and caution. By nature, they compress reality. They amputate. They omit. They simplify.

And that, of course, is the precise source of their power.

Why do scientific papers have abstracts? Why do news articles have headlines? Why do action movies have trailers cataloguing all the best explosions and postexplosion quips? It's because simplification is vital work. Nobody's got time to admire the kaleidoscopic brilliance of reality all day. We've got places to go, articles to skim, YouTube videos upon which to cast our mindless gaze. If I'm going to a new city in July, I don't seek a novelistic portrait of the humidity and heat; I look up the temperature. Such a statistic is not "living" or "deep" or "configural" (whatever that means). It's plain and clear and useful. By condensing the world, statistics give us a chance to grasp it.

And they do more, too. Statistics classify, extrapolate, and predict, allowing us to build powerful models of reality. Yes, the whole process depends on simplification. And yes, simplification is lying by omission. But at its best, statistics are an honest kind of lying. The process calls upon all the virtues of human thought, from curiosity to compassion.

In that way, statistics are not so different from stick figures. They're bad drawings of reality, missing hands and noses—yet speaking a peculiar truth all their own.

Chapter 16

WHY NOT TO TRUST STATISTICS

AND WHY TO USE THEM ANYWAY

Okay, let's get this out of our system. Statistics are lies. Not to be trusted. All of history's wittiest people have said so. Haven't they?

FAMOUS QUOTE I FOUND BY GOOGLING	REALITY I FOUND BY DEEPER GOOGLING
"There are three kinds of lies: lies, damned lies, and statistics," said Mark Twain.	People today often misattribute this to Twain, which is perhaps fair, since Twain himself misattributed it to Benjamin Disraeli. The correct attribution is unknown.
"Do not trust any statistics you did not fake yourself," said Winston Churchill. (Or perhaps: "I only believe in statistics that I doctored myself.")	A lie meant to slander Churchill. It may have originated with Nazi propagandist Joseph Goebbels.
"87% of statistics are made up on the spot."	"And 87% of quotes are misattributed on the spot," said Oscar Wilde.
"There are two kinds of statistics, the kind you look up and the kind you make up," said Rex Stout.	Rex Stout was a novelist. He didn't say this; one of his characters did.

"Politicians use statistics in the same way that a drunk uses lampposts—for support rather than illumination," said Andrew Lang.	Okay, this one is real, and it's great.
"Statistical thinking will one day be as necessary for efficient citizenship as the ability to read or write," said H. G. Wells.	Ack—even my *prostats* saying is misquoted. Wells actually said that the day is foreseeable when it will be "as necessary to be able to compute, to think in averages and maxima and minima, as it is now to be able to read and write."

What's my point? Yes, numbers can deceive. But so can words—not to mention pictures, hand gestures, hip-hop musicals, and fundraising emails. Our moral system blames the liar, not the medium chosen for the lie.

To me, the most interesting critiques of stats drive not at the dishonesty of the statistician but at the mathematics itself. We can boost statistics' value by understanding their imperfections, by seeing what each statistic aims to capture—and what it deliberately omits. Maybe then we can become the citizens that H. G. Wells envisioned.

1. THE MEAN

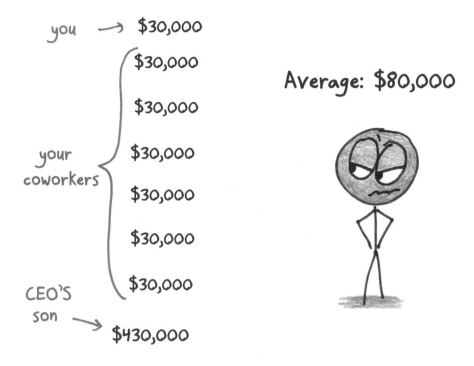

you → $30,000

$30,000

$30,000

your coworkers {
$30,000

$30,000

$30,000

$30,000

CEO'S son → $430,000

Average: $80,000

How It Works: Add up all the data you've got. Divide the total by the size of the data set.

When to Use It: The mean (or "average," as it's often known) fills a basic need in statistics: to capture the "central tendency" of a group. How tall is that basketball team? How many ice-cream cones do you sell per day? How did the class do on the test? If you're trying to summarize a whole population with a single value, then the mean is a sensible first port of call.

Why Not to Trust It: The mean considers only two pieces of information: the total, and the number of people contributing to that total.

If you've ever split a hoard of pirate treasure, then you see the danger: There are many ways to share a haul. How much does each individual contribute? Is it balanced or grossly one-sided? If I eat a whole pizza and leave nothing for you, is it fair to claim that our "average consumption" was half a pizza? You can tell your dinner guests that the "average human" has one ovary and one testicle, but will this not bring conversation to an awkward standstill? (I've tried; it does.)

Humans care about questions of allocation. The mean shrugs them off.

Mean: 3 coins. Mean: 3 coins.

There's a saving grace, which is that this feature makes the mean simple to calculate. Say your test scores are 87, 88, and 96. (Yeah, you're crushing this class.) What's your average? Don't overheat your neurons with addition and division; instead, reallocate. Take six points from your last test; give three to the first and two to the second. Now your scores are 90, 90, and 90, with a single point left over. Sharing that lonely point among the three tests, you arrive at a mean of $90\frac{1}{3}$, with no brain strain required.

2. THE MEDIAN

So, why should I invest with you?

Well, not to brag, but my fund has a median gain of 8% per year.

How It Works: The median is the middlemost member of your data set. Half your data lies below it, and half above.

When to Use It: The median, like the mean, captures the central tendency of a population. The difference lies in its sensitivity to outliers—or rather, its *in*sensitivity.

Take household income. In the US, a wealthy family may earn dozens (even hundreds) of times as much as a poor one. The mean, pretending that every family has an equal share of the total earnings, is seduced by these exceptional values, and is led away from the bulk of data points. It arrives at the figure of $75,000.

The median resists the outliers' gravitation. It names instead the nation's absolute middle household income, the perfect midpoint where half of families are wealthier and half poorer. In the US, it's near $58,000. Unlike the mean, this offers a clean picture of the "typical" household.

Why Not to Trust It: Once you've found your median, you know that half of the data lies above and half below. But how far away are those points—a hair's breadth from the median or a transcontinental flight? You're looking only at the middle slice of pie, never mind how big or small the other slices are. This can lead you astray.

When a venture capitalist invests in new businesses, she expects most of them to fail. The rare 1-in-10 home run compensates for all those little losses. But the median misses this dynamic. "The typical outcome is negative," it squawks. "Abort mission!"

Meanwhile, an insurance company builds a careful portfolio, knowing that the rare 1-in-1000 disaster can wipe out years of modest profits. But the median overlooks the potential for calamity. "Hey, the typical outcome is positive," it cheers. "Never stop!"

This is why you'll often find the median displayed alongside the mean. The median reports on the typical value; the mean, on the total. Like two imperfect witnesses, they tell a fuller story together than either can alone.

3. THE MODE

Score Category	Number of Tests
90s	0
80s	0
70s	2
60s	1
50s	1
40s	1
30s	1
20s	1

please don't ask about the mean...

How It Works: It's the most common value, the hippest, most fashionable data point.

What if each value is unique, with no repetitions? In that case, you can group the data into categories, and call the most common one "the modal category."

When to Use It: The mode shines in conducting polls and in tabulating nonnumerical data. If you want to summarize people's favorite colors, you can't very well "total the colors up" to compute a mean. Or, if you're running an election, you'll drive citizens mad if you line up the votes from "most liberal" to "most conservative" and award the office to the recipient of the median ballot.

Why Not to Trust It: The median ignores the total. The mean ignores its allocation. And the mode? Well, it ignores the total, its allocation, and just about everything else.

The mode seeks a single most common value. But "common" does not mean "representative." The modal salary in the United States is zero—not because most Americans are broke and jobless, but because wage earners are spread across a spectrum from $1 to $100 million, whereas all wageless people share the same number. The statistic doesn't illuminate anything about the US. It's true in virtually every country, an artifact of how money works.

Turning to "modal categories" can only partway solve the problem. It places a surprising power in the hands of the person presenting the data, who can gerrymander the category boundaries to suit their agenda. Depending on how I draw the lines, I can claim that the modal household in the US earns $10,000 to $20,000 (going by increments of 10,000), or $20,000 to $40,000 (going by increments of 20,000), or $38,000 to $92,000 (going by tax brackets).

Same data set, same statistic. And yet the portrait changes completely, depending on the artist's choice of frame.

4. THE PERCENTILE

Outcome	Likelihood
+ $100	90%
− $10	9%
− $1,000,000	1%

How It Works: Recall that the median splits your data set right down the middle. Well, the percentile is a median with a dimmer switch. The 50th percentile is the median itself (half the data above and half below).

But you can pick other percentiles, too. The 90th lands near the top of your data: Only 10% sits above while 90% sits below. Meanwhile, the 3rd resides near the bottom of your data set: only 3% is below, with 97% above.

When to Use It: Percentiles are handy, flexible, and perfect for that favorite human pastime: ranking things. That's why standardized tests often deliver their scores as percentiles. A raw score like "I got 72% of questions right" isn't very informative. Were these vicious snakebite questions or slow-pitch softballs? However, "I'm at the 80th percentile" reveals right where you stand. You did better than 80% of test takers and worse than 20%.

Why Not to Trust It: Percentiles share the median's Achilles heel. They tell you how much data lies above or below a certain point but not how far away it is.

Take the finance industry, which employs percentiles to measure the downside of an investment. You first imagine the spread of possible outcomes, from triumph to cataclysm, and then pick a percentile (usually the 5th), which you call "value at risk," or VaR. It aims to capture the worst-case scenario. But in fact, you'll do worse exactly 5% of the time. VaR gives no clue of how *much* worse, whether it's pennies of extra losses, or billions.

You can better visualize the landscape of possibilities by checking more VaR percentiles—for example, 3, 1, and 0.1—but by nature, a percentile cannot detect the most violent and extreme losses, and so the true worst-case scenario will always lurk just beyond sight.

5. PERCENTAGE CHANGE

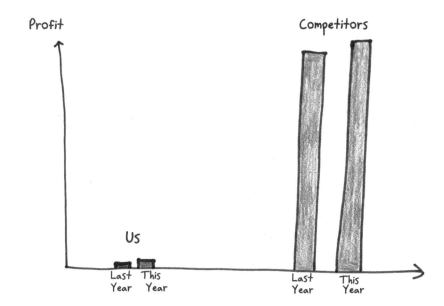

How It Works: Don't just report a change; first, divide it by the original total.

When to Use It: Percentage change is all about putting things in perspective. It frames gains and losses as a proportion of the whole.

Take a gain of $100. If I only had $200 to begin with, then this windfall marks 50% growth, and I'm dancing a hallelujah Snoopy dance. But if I already had $20,000, then my new income represents just 0.5% growth; I'll content myself with a single triumphant fist pump.

Such perspective taking is crucial when you're watching a quantity grow over time. If the Americans of 70 years ago had heard that our GDP increased by $500 billion last year, they'd be awed. If they'd heard that it grew by 3%, they'd feel right at home.

Why Not to Trust It: Hey, I'm all about perspective. But percentage change, in its effort to provide context, can annihilate it instead.

When I lived in the UK, the tasty £2-per-jar tomato sauce sometimes went on sale for £1. Those days felt like jackpots: 50% savings! I'd lug home a dozen jars, enough for months of ravioli. Then, later, I'd find myself buying plane tickets to attend a wedding in the US. Delaying the purchase a week might incur a price bump of 5%. "Ah well," I'd say as I accepted the higher rate. "It's only a little more."

Yes! 90% discount on buttons and loose sheets of paper!

An extra 1% on rent? Well, not ideal, but it's just 1%...

You can see the problem: my instincts run penny-wise and pound-foolish. The "huge" discount on tomato sauce saved me £12 while the "modest" increase on plane tickets cost me £30. A dollar is a dollar is a dollar, whether it's part of a $20 grocery bill or a $200,000 mortgage. Big price drops on cheap products are dwarfed by a small bump on a rare extravagance.

6. THE RANGE

Our students come from a wide range of socioeconomic backgrounds...

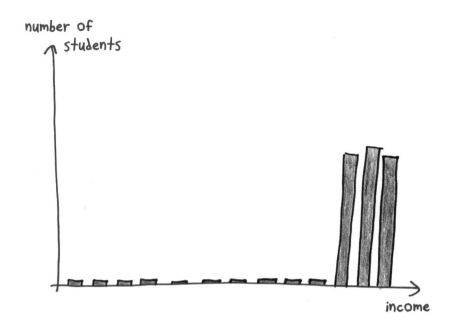

How It Works: It's the distance between the biggest and the smallest data point.

When to Use It: The mean, median, and mode deal with "central tendency": they aim to collapse a diverse population down to a single representative value. The "range" takes the opposite goal: not to sweep disagreement under the rug, but to quantify and display it, to give a measure of the data's "spread."

The range's merit lies in its simplicity. It conceives of a population as a spectrum running from "smallest" to "largest," and it tells us the width of that spectrum. It's a quick-and-dirty summary of variety.

Why Not to Trust It: The range cares only about the biggest and smallest slice of cake. That omits an awful lot of crucial information: namely, the sizes of all those middle slices. Are they close to the maximum? Close to the minimum? Scattered evenly in between? The range neither knows nor cares to ask.

The larger the data set, the more dubious the range becomes, as it disregards millions of intermediate values to inquire about the two most extreme outliers. If you were an alien who learned that there's a 7-foot range of adult human heights (from the record low of less than 2 feet to the high of almost 9), you'd be quite disappointed to visit Earth and meet all of us dull 5-to-6-footers.

7. THE VARIANCE
(AND THE STANDARD DEVIATION)

How It Works: Standard deviation tells you, in rough terms, how far the typical data point is from the mean.

If you want to cook up a variance in your home kitchen, then the recipe goes like this: (1) find the mean of your data set; (2) find how far each data point is from the mean; (3) square those distances; and (4) average those squared distances. This gives you the "average squared distance from the mean," i.e., the variance.

Attendance at My One-Man Shows	Distance from Mean	Squared Distance from Mean	Variance:
mean: 7	3 → 4	square → 16	
	4 → 3	9	
	6 → 1	1	average → 18
	7 → 0	0	
	15 → 8	64	

If you take a square root at the end, you'll get the "standard deviation." It's the more intuitive measure, since variance has strange squared units. (What's a "square dollar"? No one knows.)

Since variance and standard deviation go together, that's how I'll discuss them.

When to Use It: Like the range, variance and standard deviation quantify a data set's variety, but—and I say this with the impartiality of a loving parent—they're better. The range is a quick, makeshift measure of spread; the variance is a load-bearing pillar of statistics. By taking a contribution from every member of the data set, the variance achieves the sophistication of a symphony, in contrast to the range's two-note ditty.

The variance's logic, though convoluted, makes sense on inspection. It's all about the data's distance from the mean. "High variance" means that the data is scattered far and wide; "low variance" means that it's huddled close together.

Why Not to Trust It: Sure, the variance takes a contribution from every data point. But you can't tell *who* is contributing *what*.

In particular, a single outlier can give the variance a huge boost. Thanks to the squaring step, a single large distance (e.g., $12^2 = 144$) can yield a greater contribution than a dozen small ones (e.g., $3^2 = 9$; 12 such terms total to just 96).

Variance has another feature that stumps lots of folks. (It's not a bad trait, just a counterintuitive one.) Students tend to consider a data set with many different values (e.g., 1, 2, 3, 4, 5, 6) as more "spread out" than a data set with repeated values (e.g., 1, 1, 1, 6, 6, 6). But variance isn't interested in "variety"; it cares only about distance from the mean.

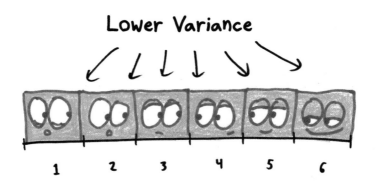

Lower Variance

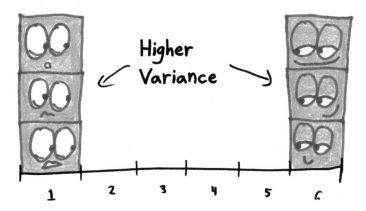

Higher Variance

In variance's eyes, the spread of the latter set (with its repeated far-from-the-mean values) outweighs the spread of the former (with its nonrepeated but closer-to-the-mean values).

8. THE CORRELATION COEFFICIENT

Try our energy drink—it's highly correlated with performance!

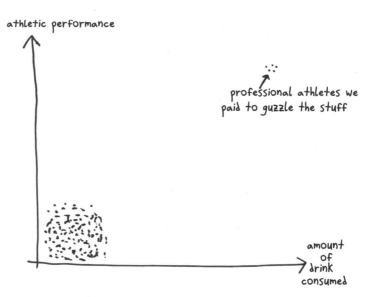

How It Works: A correlation measures the relationship between two variables. A person's height and weight, for example. Or a car's price and its number sold. Or a movie's budget and box office receipts.

The scale runs from a maximum of 1 ("wow, they totally go together") to a midpoint of zero ("huh, there's no connection here") to a minimum of -1 ("hmm, they go *opposite* directions").

That's the quick-hit summary, anyway. To see how the correlation coefficient *actually* works, check the endnotes.

When to Use It: Are wealthier nations happier? Does "broken windows" policing prevent crime? Does drinking red wine prolong life spans, or just dinner parties? All these questions deal with connections between pairs of variables, between imagined causes and conjectured effects. Ideally, you'd answer them by running experiments. Give 100 people red wine, give 100 grape juice, and see who lives longer. But such research is slow, expensive, and often unethical. Think of that piteous wine-deprived control group.

Correlation lets us tackle the same question sideways. Find a bunch of people, measure their wine intakes and life spans, and see if the wine drinkers live longer. Admittedly, even a strong correlation can't determine causation. Maybe wine lengthens life. Maybe long life drives people to drink. Maybe both are driven by some third variable (e.g., rich people both live longer and can afford more wine). It's impossible to know.

Even so, correlational studies offer a great starting point. They're cheap and fast, and they allow for large data sets. They can't pinpoint causes, but they can offer tantalizing clues.

Why Not to Trust It: The correlation coefficient is one of the most aggressive statistical summaries of all. It collapses hundreds or thousands of data points, each with two measured variables, into a single number between -1 and 1. Suffice it to say, that's going to leave some things out, a fact illustrated by the mathematical oddity known Anscombe's quartet.

Let us step into Anscombe Academy for Witchcraft and Wizardry, where students have spent weeks preparing for exams in four classes: Potions, Transfiguration, Charms, and Defense Against the Dark Arts. For each exam, we shall consider two variables: the hours each student spent studying and that student's score (out of 13) on the exam.

From the summary stats, you'd think the four tests were identical:

Average <u>hours</u> <u>studying</u>: **9**

<u>Variance</u> in <u>hours</u> studying: **11**

<u>Average</u> <u>score</u>: **7.5** (to nearest 0.01)

<u>Variance</u> in <u>score</u>: **4.125** (to nearest 0.125)

<u>Correlation</u>: **0.816** (to nearest 0.001)

And yet . . . well, just look. (Each dot represents a student.)

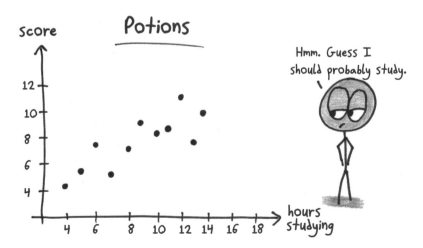

The first, Potions, fits my stereotype of how tests work. Studying more tends to improve your result, but it's no sure thing. Random noise intervenes. Thus, the correlation is 0.816.

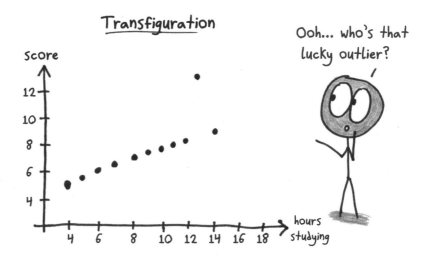

The Transfiguration scores, meanwhile, follow a perfect linear relationship, with every extra hour of study yielding an extra 0.35 points on the test—except for one peculiar outlier child, who lowers the correlation from a perfect 1 down to 0.816.

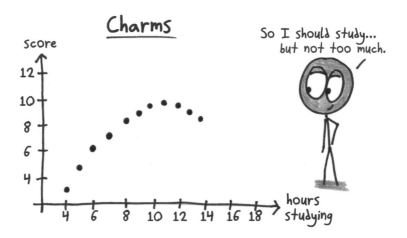

The Charms test follows an even more certain pattern: studying improves your scores, but with diminishing marginal returns. When you reach 10 hours, extra study time begins to harm your score (perhaps by costing you sleep). However, correlation is built to detect linear relationships, and so it misses the nature of this quadratic pattern, resulting in a coefficient of 0.816.

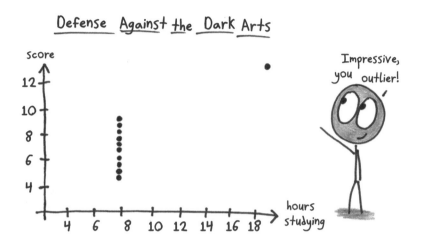

Finally, in Defense Against the Dark Arts, every pupil studied for eight hours, meaning that study time can't help you predict the final score. There's just one exception: a hardworking outlier whose 19 hours of study paid off with the runaway best score. That single data point suffices to strengthen the correlation from 0 all the way up to . . .

0.816.

Each test follows its own distinct logic, obeying its own unique pattern. But the correlation coefficient misses the story.

Then again, that's the nature of statistics. As I like to say:

A statistic is an imperfect witness. It tells the truth, but never the whole truth.

Feel free to quote me. Or, if you favor the traditional method, just make up your own saying and attribute it to me.

Chapter 17

THE LAST .400 HITTER

THE RISE AND FALL OF BATTING AVERAGE

E ver since its birth, baseball has been a game of numbers. Today, Wikipedia lists 122 types of baseball stats, from DICE to FIP to VORP, and I suspect these only scratch the surface. Go ahead: type three random letters; I wager that someone, somewhere, keeps a fastidious record of that statistic.

ILL–innings lost to lollygagging

ETV–enjoyability on television

CTD–catch, throw, dance

VHW–very high waistbands

NGS–noogies

BFD–best friends on defense

Time to cut the ILLs, boys. Next time your VHW brings down our team's ETV, I'm gonna give you so many NGS you won't know your CTDs from your BFDs.

This chapter will explore a single statistic, from its humble origin to its gradual demise. That statistic is BA: Boston accents. Sorry, BA: beer absorbed. Okay, fine, it's BA: batting average.

Once upon a time, batting average reigned supreme. These days, stat-isticians treat it like a piece of kitsch, a relic of simpleminded days. Is it time to put the batting average out to pasture? Or does the old veteran with the aching joints still carry a spark of magic?

1. BOXED LIGHTNING

In 1856, an Englishman named Henry Chadwick, working as a cricket reporter for the *New York Times*, happened to attend his first baseball game. It caught his fancy. "In baseball all is lightning," he gushed, as only a cricket fan could. Like a sloth smitten with the flash and pizzazz of a tortoise, Chadwick soon devoted his life to this American pastime. He served on rules committees, wrote the sport's first book, and edited its first annual guide. But what earned Chadwick his nickname as "the Father of Baseball" was something more elemental: the stats.

Chadwick invented the "box score," a table for tracking the game's key events. By scanning columns of numbers—runs, hits, putouts, and so on—you could almost watch the innings unfold. Box scores weren't about long-term predictive power or statistical validity. They were about credit and blame, heroes and villains, a story told in numbers. They summarized weather conditions and highlighted key events, aiming (in a time before radio, action photography, or MLB.com) to help you glimpse the action. It was *SportsCenter* for the 1870s.

Chadwick's idea for "batting average" came from cricket, where there are only two bases and you score a run every time you move from one to the other. Cricketers keep batting until they make an out, by which time a good player will have scored dozens of runs. (The all-time record is 400.)

Thus, cricket's batting average is defined as "runs scored per out made." A great player can maintain an average of 50, even 60.

This definition made little sense in baseball, where a single hit ends your at-bat. So, like any good mathematician, Chadwick played with the rules, trying out a few definitions before settling on the one used today.

$$BA = \frac{\text{bats used}}{\text{games played}}$$

hmm...

$$BA = \frac{\text{Batman villains defeated}}{\text{Batman villains confronted}}$$

nO...

$$BA = \text{an arbitrary number capturing the statistician's subjective emotional state}$$

well...

$$BA = \frac{\text{hits}}{\text{at-bats}}$$
(not counting walks)

That's it!

Batting average aims to capture your rate of success with a simple fraction: your number of successful hits divided by the number of hits plus the number of outs you make. Chadwick called it the "one true criterion of skill at the bat."

In theory, batting averages range from .000 (never hitting) to 1.000 (never failing to hit). In practice, almost all players cluster between .200 and .350. That's not much of a gap. Baseball's royals (hitting 30% of the time) and its peasants (hitting 27.5%) differ by just one hit in every 40 attempts. The naked eye can't tell the difference. Even over a whole season, a "worse" hitter might outshine a "better" one by sheer luck.

Will the .300 hitter have more hits than the .275 hitter?

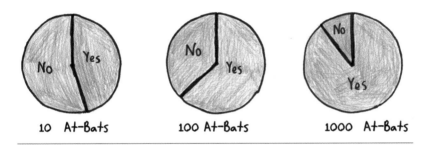

| 10 At-Bats | 100 At-Bats | 1000 At-Bats |

Wait... even _two_ seasons might not be enough to show I'm better?!

I think of it as, two seasons might not be enough for you to _be_ better.

0.300 Hitter

0.275 Hitter

Hence, the stats. A player's batting average is like a stop-motion video of a flower growing from seedling to full bloom. It reveals a truth that otherwise would lie beyond our senses. Instead of lightning in a bottle, we've got lightning in a box score.

Statistics, like probability, bridges two worlds. First, there's the messy day-to-day reality of bad hops and lucky bounces. Second, there's the long-run paradise of smooth averages and stable tendencies. Probability begins with the long-run world, and imagines how things might turn out on a single day. Statistics does the opposite: it begins with the day-to-day mess, and strives to infer the invisible long-run distribution from which that data emerged.

Put another way: a probabilist starts with a deck of cards, and describes what hands might come up. A statistician looks at the hands on the table, and tries to infer the nature of the deck.

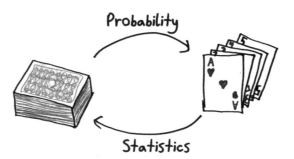

Baseball, perhaps unique among sports, deals you enough cards to have a real shot. In every 162-game season, a team's batters face roughly 24,000 pitches. For soccer to offer the same wealth of discrete data points, you'd have to reset the ball at center field every five seconds, all season long. Better yet, whereas other team sports unfold in multiperson melees, each baseball player bats alone, his data clean and separable from his teammates'.

That's the glory of batting average. But every statistic leaves something out—and in this case, what's missing is something crucial.

2. THE OLD MAN AND THE OBP

In 1952, *Life* magazine published the first edition of Ernest Hemingway's *The Old Man and the Sea*. The issue sold 5 million copies; the novelist won a Nobel Prize.

On August 2, 1954, *Life* chose to steer the national conversation a different direction: baseball statistics. Under the headline "Goodby to Some Old Baseball Ideas," Pittsburgh Pirates general manager Branch Rickey posited an equation that would take 10 pages to unpack:

$$\left(\frac{H+BB+HP}{AB+BB+HP} + \frac{3(TB-H)}{4AB} + \frac{R}{H+BB+HP}\right) - \left(\frac{H}{AB} + \frac{BB+HB}{AB+BB+HB} + \frac{ER}{H+BB+HB} - \frac{SO}{8(AB+BB+HB)} - F\right) = G$$

See? Clear now?

Why... it's as clear as the infield fly rule...

The formula itself was scarcely grammatical. The equals sign doesn't really mean "equals," nor is the minus sign actual subtraction. Still, the article gave a penetrating critique of several "old baseball ideas"—batting average chief among them. The attack (credited to Rickey but ghostwritten by Canadian statistician Allan Roth) begins with two letters: BB, for "base on balls." Or, in common parlance, a walk.

As baseball came of age in the 1850s, a player batted until he either hit the ball into play, or swung and missed three times. With patient enough batters, the game flowed like cold molasses. Thus, in 1858, "called strikes" were born. If a player let a juicy pitch sail past, then it was treated as equivalent to swinging and missing. But now the pendulum swung too far; cautious pitchers refused to throw anything hittable. The solution, introduced in 1863, was to also call "balls": pitches deemed too off target for the batter to hit. Enough such "balls" would grant the batter a free walk to first base.

Walks stumped Chadwick. Cricket's closest equivalent is a "wide," generally viewed as a mistake by the thrower. So batting average ignored walks, as if the at-bats had never occurred. Walks weren't deemed an official statistic until 1910.

Today, the most skilled and patient hitters walk 18% or 19% of the time; their reckless, swing-happy peers, just 2% or 3%. Hence the first term in Rickey's equation: a convoluted expression for what we now call "on-base percentage," or OBP. It's your rate of getting on base, via either hits or walks—in other words, your rate of *not* making an out.

Which statistic better predicts the number of runs a team will score: BA or OBP? Running the correlations for 2017, BA is pretty strong, with a coefficient of 0.73. But OBP is outstanding, with a coefficient of 0.91.

BA is good.

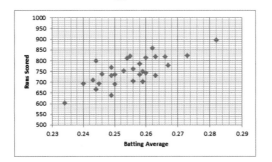

Good correlation.

OBP is better.

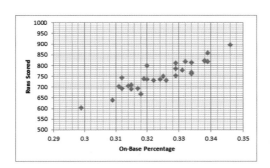

Great correlation!

(each dot represents a team's 2017 season)

Next, Rickey (i.e., Roth) highlighted another shortcoming of batting average. Hits come in four flavors, from a single (one base) to a home run (four bases). As with desserts or retweets, more is better—but batting average treated them all the same. Hence, the second term in Rickey's equation, which amounts to a measure of "extra bases" beyond first.

Today, we prefer a related statistic: "slugging percentage," or SLG. It calculates the average number of bases achieved per at-bat, from a minimum of 0.000 to a theoretical maximum of 4.000 (a home run every time). In practice, no hitter has ever topped 1.000 for a whole season.

Like batting average, SLG ignores walks, and like any stat, it collapses meaningful distinctions. For example, to slug .800 over 15 at-bats, you need to get 12 total bases (since 12/15 = 0.8). There are a lot of ways to do that, not all of them equal in value:

homer, out, out, out, out
homer, out, out, out, out
homer, out, out, out, out

3 homers x 4 bases each
= 12 bases

out, triple, out, out, out
triple, out, out, triple, out
out, out, triple, out, out

4 triples x 3 bases each
= 12 bases

out, out, double, out, double
out, double, out, double, out,
double, out, double, out, out,

6 doubles x 2 bases each
= 12 bases

single, single, single, single, out
single, single, single, single, out
single, single, single, single, out

12 singles x 2 bases each
= 12 bases

Since OBP and SLG each shed light on a different corner of the game, they're often used in conjunction. One common shortcut is to add the two, arriving at a statistic known as "on-base plus slugging," or OPS. In the 2017 data, its correlation with runs scored is an eye-popping 0.935—better than either OBP or SLG alone.

On the 50th anniversary of the *Life* article, the *New York Times* showed the formula to New York Yankees general manager Brian Cashman. "Wow!" he said. "The guy was generations ahead of his time." His praise belies the damning truth: Cashman had never heard of the *Life* piece. Even after its publication, batting average would reign for decades more, while OBP and SLG huddled in the shadows. The baseball dialogue in *The Old Man and the Sea* probably had more impact on the game than Rickey's manifesto.

What was baseball waiting for?

3. KNOWLEDGE MOVES THE CURVE

To revolutionize anything, you need two conditions: the knowledge and the need.

The knowledge came, in large part, from writer Bill James. In 1977, while working night shifts as a security guard, he self-published the first *Bill James Baseball Abstract*, a peculiar 68-page document consisting mostly of statistics, with rigorous answers to insider questions like "Which pitchers and catchers allow the most stolen bases?" It sold just 75 copies, but the buzz was good. The next year's edition sold 250. Five years later, James landed a major publishing deal. In 2006, *Time* magazine profiled James (now on the payroll of the Boston Red Sox) in its "Time 100" as one of the most influential people on planet Earth.

James's clear-eyed analytical approach sparked a statistical renaissance in baseball. He dubbed it "sabermetrics." Among the movement's insights was that batting average is a sideshow, a crude indicator for an actual outcome—like inferring the quality of a dinner by sampling one ingredient. If you're serious about assessing the meal, you need to taste *all* the ingredients—or, better yet, the dish itself.

As the archivists at *Life* magazine can attest, this knowledge was old. What brought it into the forefront was not just James, but also a need born from the changing economic conditions of baseball. Until the early 1970s, baseball players lived under the shadow of the "reserve clause." It meant that, even after a contract expired, the team retained rights to the player. You couldn't sign (or, for that matter, negotiate) anywhere else, except with your old club's permission.

Then, in 1975, an arbiter's ruling redefined the reserve clause, launching the era of "free agency." With the floodgates open, salaries soared.

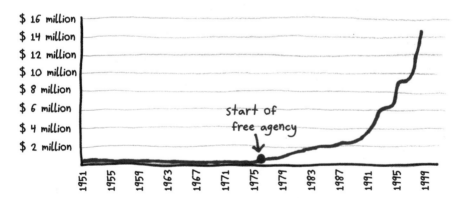

A decade earlier, owners could buy players like groceries. Now, the groceries had agents, and those agents had salary demands. The new financial pressure should have nudged owners away from crude measures like BA toward more reliable stats like OBP and SLG—but as everyone but Henry Chadwick knows, baseball is a slow-moving game. It took 20 years for one team, the Oakland A's, to gain sentience and begin using OBP to evaluate players.

The fire started under general manager Sandy Alderson in the early 1990s and burned bright under his successor, Billy Beane. Soon, the A's built an outrageous run of success on smart statistical shopping. Then, in 2003, a Bay Area resident named Michael Lewis wrote a book about Billy Beane. He called it *Moneyball*, and along the way to selling 67 quadrillion copies, it did what *Life* magazine had failed to do: it bid "goodby" to some old baseball ideas. With Lewis's help, OBP and SLG leapt from the fringes of fandom to the front-office mainstream.

4. THE DRAMA OF THE 4TH DECIMAL

"Baseball is the only field of endeavor," outfielder Ted Williams once said, "where a man can succeed three times out of ten and be considered a good performer."

In 1941, Williams was on pace to do one better: to succeed *four* times out of 10. This is called "batting .400" and it makes you a demigod, a legend. Williams entered the final week of the season at .406, poised to become the first .400 hitter in 11 years.

Then he faltered. In the next four games, he got just three hits in 14 chances, bringing his average down to a heartbreaking .39955.

The number looks fake, like an example you'd cook up to test a student's grasp on decimals. Is it .400, or not? The next day, leading newspapers made clear their view: It was not. "Williams at .3996," the *New York Times* said. "Williams Drops Under .400," announced the *Chicago Tribune*. "Williams Falls to .399," the *Philadelphia Inquirer* reported, bending the rules of rounding, while Williams's hometown paper, the *Boston Globe* echoed that stat: "Average Now Is .399."

How can you resist a sport that musters such passion for the fourth decimal place?

THE GREAT ROUNDING CONTROVERSY OF 1941

The final two games of the season fell back-to-back on September 28. The evening before, Williams prowled 10 miles across the streets of Philadelphia, unable to sleep. Prior to the first game, according to one sportswriter, he "sat on the bench, biting his fingernails. His mammoth hands trembled." He later reported "shaking like a leaf when [he] went to bat the first time."

But the 23-year-old persevered. That afternoon, he managed six hits in eight chances, boosting his average to .4057. (The headline writers didn't quibble about calling this .406.) Nearly 80 years have elapsed, and no one has hit .400 since.

In 1856, Henry Chadwick glimpsed a dusty, bare-handed game, an excuse for a summer afternoon. The Englishman gave it numbers. Numbers gave it reach. A century and a half later, baseball is a corporate triumph, with team payrolls running in the hundreds of millions. The 19th-century batting average struggles to keep pace in a 21st-century game, like a boy trying to field a scorching line drive with his naked hands.

Still, against all the odds, .400 retains its magic. In April and May, when the season is young and the sample sizes are as small as spring buds, you can often find a player or two flirting with the .400 line. Before long, they'll fade. But for a week or so there's a breath of hope across the land, a sense that mythical creatures like dragons and .400 hitters have not yet vanished. The .500 OBP and the .800 SLG will never stir the heart in quite the same way. We love .400 neither for its predictive power nor for its mathematical elegance, but because it carries an electric charge, because it tells a story in three decimal places—four, if you're feeling picky.

Maybe no one will ever hit .400 again. Maybe it will happen next year. For his part, Williams shrugged it all off. "If I had known hitting .400 was going to be such a big deal," he said 50 years later, "I would have done it again."

Chapter 18

BARBARIANS AT THE GATE OF SCIENCE

THE P-VALUE CRISIS

Hooray! It's fun science fact time!

To begin: Did you know that you're more likely to cheat on a test after reading a passage that argues there's no such thing as free will?

Or that after plotting two close-together points on graph paper, you feel emotionally closer to your family members than after plotting two far-apart points?

Or that adopting certain "power poses" can inhibit stress hormones while boosting testosterone, prompting others to judge you as more confident and impressive?

I'm not making these up. They're real studies, performed by real scientists, wearing real lab coats and/or jeans. They're grounded in theory, tested by experiment, and vetted by peers. The researchers obeyed the scientific method and hid no rabbits up their sleeves.

Yet these three studies—and dozens more, in fields as far-flung as marketing and medicine—are now in question. They may be wrong.

Across the sciences, we are living through an epoch of crisis. After decades of the best science they could muster, many scholars now find their life's work hanging in the balance. The culprit isn't dishonesty, or a lack of integrity, or too many passages arguing against free will. The illness runs deeper, all the way down to a single statistic at the heart of the research process. It's a figure that made modern science possible—and that now threatens its stability.

1. TO CATCH A FLUKE

Every science experiment asks a question. Are gravitational waves real? Do millennials hate solvency? Can this new drug cure antivax paranoia? No matter the question, there are two possible truths ("yes" and "no") and, given the inherent unreliability of evidence, two possible outcomes ("you get it right" and "you get it wrong"). Thus, experimental results can fall into four categories:

Question: Are there ghosts?

The data say "yes!"
They're onto us!
True Positive

The data say "no."
Srsly?
False Negative

The data say "yes!"
False positive

The data say "no."
True Negative

Scientists want **true positives**. They are known as "discoveries" and can win you things like Nobel Prizes, smooches from your romantic partner, and continued funding.

True negatives are less fun. They're like thinking you'd tidied the house and done the laundry, only to realize that, nope, that was just in your head. You'd rather know the truth, but you wish it were otherwise.

By contrast, **false negatives** are haunting. They're like looking for your lost keys in the right place but somehow not seeing them. You'll never know how close you were.

Last is the scariest category of all: **false positives**. They are, in a word, "flukes," falsehoods that, on a good hair day, pass for truths. They wreak havoc on science, sitting undetected in the research literature for years and spawning waste-of-time follow-ups. In science's never-ending quest for truth, it's impossible to avoid false positives altogether—but it's crucial to keep them to a minimum.

That's where the p-value comes in. Its whole purpose is to filter out flukes.

To illustrate, let's run an experiment: *does eating chocolate make people happier?* At random, we divide our eager subjects into two groups. Half eat chocolate bars; half eat graham crackers. Then, all report their happiness on a scale from 1 (agony) to 5 (bliss). We predict that the chocolate group will score higher.

But there's a danger. Even if chocolate makes no difference, one group is bound to score a little higher than the other. For example, look what happened when I generated five random samples from the same population:

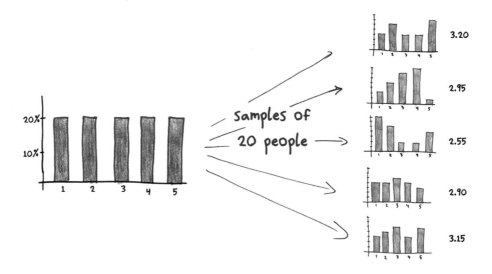

Thanks to random chance, two theoretically identical groups may deliver very different results. What if, by sheer coincidence, our chocolate group scores higher? How will we distinguish a genuine happy boost from a meaningless fluke?

To weed out flukes, the p-value incorporates three fundamental factors:

1. **The size of the difference.** A razor-thin margin (say, 3.3 vs. 3.2) is much likelier to occur by coincidence than a substantial gap is (say, 4.9 vs. 1.2).

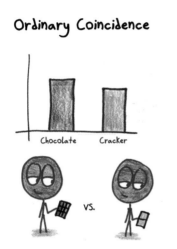

2. **How big the data set is.** A two-person sample inspires little confidence. Maybe I happened to give the chocolate to an enthusiastic lover of life, and the graham cracker to an ungrateful nihilist. But in a randomly divided sample of two *thousand* people, individual differences should wash out. Even a smallish gap (3.08 vs. 3.01) is unlikely to happen by fluke.

3. **The variance within each group.** When the scores are wild and high-variance, it's easy for two groups to get different results by fluke. But if the scores are consistent and low-variance, then even a small difference is hard to achieve by coincidence.

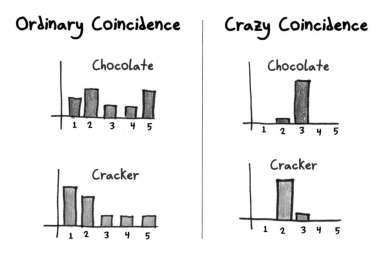

The p-value boils all of this information down into a single number between zero and 1, a sort of "coincidence craziness score." The lower the value, the crazier it would be for these results to happen by coincidence alone. A p-value near zero signals a coincidence so crazy that perhaps it isn't a coincidence at all.

(For a slightly more technical discussion, see the endnotes.)

Some p-values are easy to interpret: 0.000001 marks a one-in-a-million fluke. Such coincidences are so rare that the effect is almost certainly real: in this case, that chocolate makes people happier.

Meanwhile, a p-value of 0.5 marks a 1-in-2 event. Such results happen . . . well, half the time. They're as common as weeds. So, in this case, it would seem that chocolate makes no difference.

Between these clear-cut cases lies a disputed borderland. What about a p-value of 0.1? What about 0.01? Do these numbers mark out lucky flukes or results so extreme that they probably aren't flukes at all? The lower the p-value, the better; but how low is low enough?

2. CALIBRATING THE FLUKE FILTER

In 1925, a statistician named R. A. Fisher published a book called *Statistical Methods for Research Workers*. In it, he proposed a line in the sand: 0.05. In other words, let's filter out 19 of every 20 flukes.

Why let through the other one in 20? Well, you can set the threshold lower than 5% if you like. Fisher himself was happy to consider 2% or 1%. But this drive to avoid false positives incurs a new risk: false *negatives*. The more flukes you weed out, the more true results get caught in the filter as well.

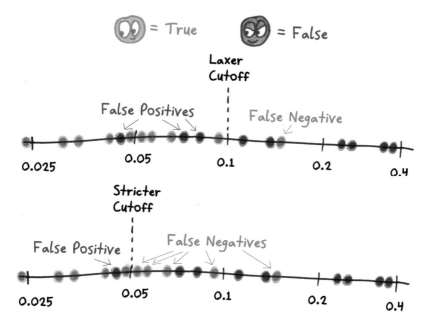

Suppose you're studying whether men are taller than women. Hint: they are. But what if your sample is a little fluky? What if you happen to pick taller-than-typical women and shorter-than-typical men, yielding an average difference of just 1 or 2 inches? Then a strict p-value threshold may reject the result as a fluke, even though it's quite genuine.

The number 0.05 represents a compromise, a middle ground between incarcerating the innocent and letting the guilty walk free.

For his part, Fisher never meant 0.05 as an ironclad rule. In his own career, he showed an impressive flexibility. Once, in a single paper, he smiled on a p-value of 0.089 ("some reason to suspect that the distribution . . . is not wholly fortuitous") yet waved off one of 0.093 ("such association, if it exists, is not strong enough to show up significantly").

To me, this makes sense. A foolish consistency is the hobgoblin of little statisticians. If you tell me that after-dinner mints cure bad breath (p = 0.04), I'm inclined to believe you. If you tell me that after-dinner mints cure osteoporosis (p = 0.04), I'm less persuaded. I admit that 4% is a low probability. But I judge it even *less* likely that science has, for decades, overlooked a powerful connection between skeletal health and Tic Tacs.

All new evidence must be weighed against existing knowledge. Not all 0.04s are created equal.

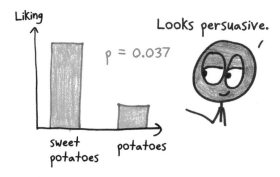

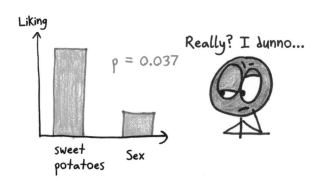

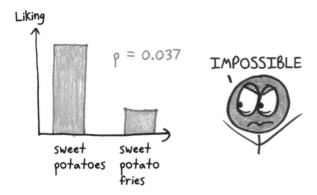

Scientists get this. But in a field that prides itself on standardization and objectivity, nuanced case-by-case judgments are hard to defend. And so, as the 20th century wore on, in human sciences like psychology and medicine the 5% line evolved from "suggestion" to "guideline" to "industry standard." $p = 0.0499$? Significant. $p = 0.0501$? Sorry, better luck next time.

Does this mean that 5% of certified results are flukes? Not quite. The reality is the reverse: 5% of flukes can become certified results. If that sounds equivalent, it's not.

It's much scarier.

Imagine the p-value as a guardian in the castle of science. It wants to welcome the true positives inside the walls while repelling the barbarian false positives at the gate. We know that 5% of barbarians will slip through, but on balance, this seems good enough.

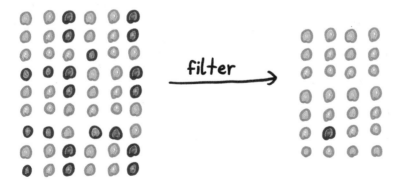

However, what if the attacking barbarians outnumber our own troops 20 to one? Five percent of the invading forces will equal the entirety of the civilized ones.

Worse yet, what if there are a *hundred* barbarians for every honest soldier? Their 5% will overwhelm the entire civilized army. The castle will teem with false positives, while the true ones cower in corners.

The danger, then, lies in scientists running too many studies where the real answer is "no." Does lip-syncing make your hair blonder? Can wearing clown shoes cause acid rain? Run a million junk studies, and 5% will clear the filter. That's 50,000. They'll flood the scientific journals, splash across headlines, and make Twitter even less readable than usual.

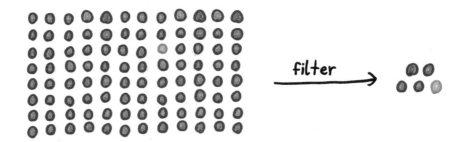

If that's not discouraging enough, it gets worse. Without meaning to, scientists have equipped the barbarians with grappling hooks and battering rams.

3. HOW FLUKES BREED

In 2006, a psychologist named Kristina Olson began to document a peculiar bias: Children prefer lucky people to unlucky ones. Olson and her collaborators found the bias across cultures, from age three to adulthood. It applied to those suffering little mishaps (like falling in the mud) or big calamities (like hurricanes). The effect was robust and persistent—a true positive.

Then, in 2008, Olson agreed to advise the senior thesis of a feckless 21-year-old named "me." With her copious help, I devised a modest follow-up, exploring whether five- and eight-year-olds would give more toys to the lucky than the unlucky.

I tested 46 kids. The answer was "no."

If anything, the reverse held: my subjects seemed to give more to the unlucky than the lucky. Far from "fun science fact time," this felt obvious; of course you give a toy to someone who lost one. Needing to squeeze 30 pages out of the experience, I looked to my data. Each subject had answered eight questions, and I had tested a variety of conditions. Thus, I could carve up the numbers in several ways:

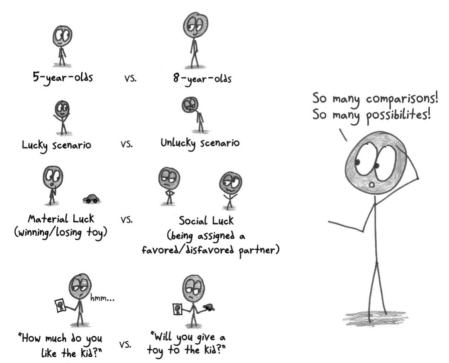

5-year-olds   VS.   8-year-olds

Lucky scenario   VS.   Unlucky scenario

Material Luck   VS.   Social Luck
(winning/losing toy)   (being assigned a
favored/disfavored partner)

"How much do you   VS.   "Will you give a
like the kid?"   toy to the kid?"

So many comparisons!
So many possibilites!

Here, in the unassuming columns of a spreadsheet, is where the danger began.

By all appearances, my thesis was a barbarian at the gates. The crucial p-value was well above 0.05. But with an open mind, other possibilities emerged. What if I considered just the five-year-olds? Or just the eight-year-olds? Or just the lucky recipients? Or just the unlucky ones? Did gender make a difference? What if eight-year-old girls proved more context-sensitive than five-year-old boys in giving to children whom they rated at least 4 on the six-point "liking" scale?

What if, what if, what if . . .

By slicing and reslicing my data, I could transform one experiment into 20. It didn't matter if the p-value rejected my barbarian once, twice, or 10 times. I could equip it with new disguises until it finally snuck through into the castle.

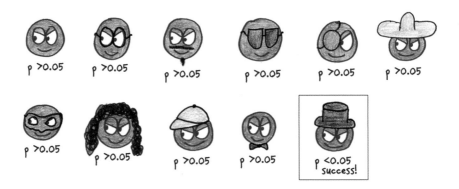

Thus is born perhaps the greatest methodological crisis of the young century: p-hacking. Take a bunch of truth-loving scientists, place them in a winner-take-all competition to achieve positive results, and then watch as, in spite of themselves, they act like 21-year-old me, rationalizing dodgy decisions. "Well, maybe I could run the numbers again . . . " "I *know* this result is real; I just need to exclude those outliers . . . " "Ooh, if I control for the 7th variable, the p-value drops to 0.03 . . . " Most research is ambiguous, with a mess of variables and a multitude of defensible ways to interpret the data. Which are you going to pick: the method that brands your results "insignificant" or the one that nudges them below 0.05?

It's not hard to dredge up this kind of nonsense. At the website Spurious Correlations, Tyler Vigen combs through thousands of variables to find pairs that share a close yet coincidental alignment. For example, the number of people who drowned by falling into a pool, from 1999 to 2009, shows a shocking correspondence with the number of films featuring Nicolas Cage.

Bombard the fluke filter, and a few false positives are bound to sneak through.

Wanting to verify this for myself, I split 90 people into three groups. Each subject got a drink: tap water, bottled water, or a mixture. Then I measured four variables from each subject: their 100-meter running time, their IQ score, their height, and their love for Beyoncé. Afterward, I looked at all possible comparisons. Were the tap-water drinkers faster runners than the bottled-water ones? Were the bottled-water folks bigger Beyoncé fans than the mixture drinkers? And so on. The research took me eight months.

Except, of course, it didn't. I simulated it in a spreadsheet, running it 50 times in a few minutes.

Every subject was identical in principle: a collection of random numbers generated by the same process. Any differences that emerged were necessarily flukes. Still, with three groups to compare, and four variables for comparison, I achieved 18 "significant" results in 50 trials.

The p-value wasn't letting through 1-in-20 flukes. It was more than 1 in 3.

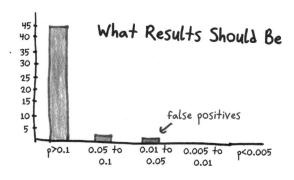

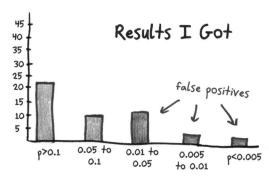

There are other ways to hack a p-value. A 2011 survey found large fractions of psychologists admitting, under the veil of anonymity, to various "questionable research practices":

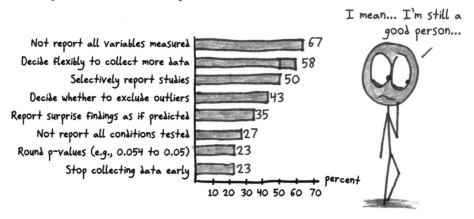

Not report all variables measured	67
Decide flexibly to collect more data	58
Selectively report studies	50
Decide whether to exclude outliers	43
Report surprise findings as if predicted	35
Not report all conditions tested	27
Round p-values (e.g., 0.054 to 0.05)	23
Stop collecting data early	23

percent
10 20 30 40 50 60 70

I mean... I'm still a good person...

Even the most innocent-looking of these dogs can bite. Like, say, gathering more data when your initial results prove inconclusive. Seems harmless, right?

To gauge the power of this p-hack, I simulated a study I call "Who's the Better Coin Flipper?" It's simple: two "people" (read: spreadsheet columns) each flip 10 coins. Then we check if one of them achieved more heads. In 20 simulations, I achieved a significant result once. That's exactly as you'd expect for p = 0.05: 20 trials, one fluke. Reassuring.

Then, I allowed myself the freedom to keep going. Flip another coin. And another. And another. Stop the study if the p-value dips below 0.05 (or if we reach 1000 flips with no success).

The results transformed. Now, 12 out of 20 studies yielded a significant result.

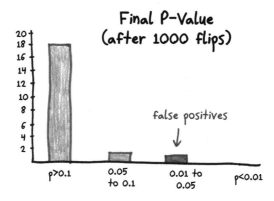

Final P-Value
(after 1000 flips)

20 18 16 14 12 10 8 6 4 2

false positives
↓

p>0.1   0.05 to 0.1   0.01 to 0.05   p<0.01

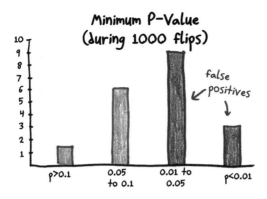

Such tricks aren't good science, but they're not quite fraud, either. In their paper reporting the survey results, the three authors called such methods "the steroids of scientific competition, artificially enhancing performance and producing a kind of arms race in which researchers who strictly play by the rules are at a competitive disadvantage."

Is there any way to level the playing field?

4. THE WAR ON FLUKES

The replication crisis has rekindled an old rivalry between two gangs of statisticians: the frequentists and the Bayesians.

Ever since Fisher, frequentists have reigned. Their statistical models aim for neutrality and minimalism. No judgment calls. No editorializing. The p-value, for instance, doesn't care whether it's testing a surefire hypothesis or a mad-scientist one. That kind of subjective analysis comes later.

Bayesians reject this impartiality. Why should statistics feign indifference between plausible hypotheses and absurd ones, as if all 0.05's are created equal?

The Bayesian alternative works something like this. You begin with a "prior": an estimate of your hypothesis's probability. "Mints cure bad breath"? High. "Mints cure bad bones"? Low. You bake this estimate into the mathematics itself, via a rule called Bayes's formula. Then, after you run the experiment, statistics help you update the prior, weighing the new evidence against your old knowledge.

Bayesians don't care whether the results clear some arbitrary fluke filter. They care whether the data suffice to persuade us, whether they move the needle on our prior beliefs.

Bayesianism: Judgment baked into statistics.

This hypothesis is ludicrous. I'd call the probability 1 in 3 million.

Experiment →

Okay, fine, the probability is 1 in 60,000. still not persuaded.

The Bayesians feel that their time has come. The frequentist regime, they assert, has led to ruin, and it's time to inaugurate a new era. Frequentists reply that priors are too arbitrary, too vulnerable to abuse. They propose their own reforms, such as lowering the p-value threshold from 0.05 (or 1 in 20) to 0.005 (or 1 in 200).

While statistics deliberates, science isn't waiting around. Researchers in psychology have begun the slow, difficult process of eliminating p-hacks. It's a litany of protransparency reforms: preregistering studies, listing every variable measured, and specifying in advance one's rules for when to stop data collection and whether to exclude outliers. All of this means that if a false positive manages to finesse its p-value below 0.05, you can glance at the paper and see the 19 failed attempts that came before. One expert told me that it's this suite of reforms, much more than questions of mathematical philosophy, that will make the difference.

For what it's worth, my senior thesis adhered to most of these standards. It listed all variables gathered, excluded no outliers, and made clear the exploratory nature of the analysis. Still, when I showed Kristina this chapter, she noted, "It's really funny in 2018 to read about a thesis from 2009. Today my thesis students all preregister their hypotheses, sample sizes, etc. How much we have grown and learned!"

All of this should slow the stream of barbarians through the gates. Still, that leaves the matter of the barbarians already inside. To identify those, there is only one solution: replication.

Imagine 1000 people predict 10 coin tosses. Chances are that one of them will guess all 10 correctly. Before you start snapping selfies with your newfound psychic friend, you ought to replicate. Ask this person to predict 10 more flips. (And perhaps 10 more. And maybe 10 after that.) A true psychic should manage the feat. A lucky guesser will revert to normal.

The same is true of all positive results. If a finding is true, then rerunning the experiment should generally yield the same outcome. If it's false, then the result will vanish like a mirage.

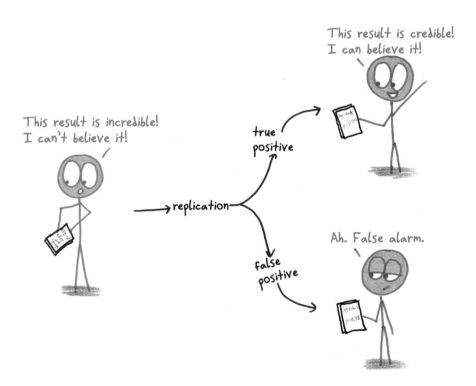

Replication is slow, unglamorous work. It takes time and money while producing nothing new or innovative. But psychology knows the stakes and is beginning to face its demons. One high-profile project, published in 2015, performed careful replications of 100 psychological studies. The findings made headlines: 61 of the 100 failed to replicate.

In that grim news, I see progress. The research community is taking a sobering look in the mirror and owning up to the truth, as ugly as it may be. Now social psychologists hope that other fields, such as medicine, will follow their lead.

Science has never been defined by infallibility or superhuman perfection. It has always been about healthy skepticism, about putting every hypothesis to the test. In this struggle, the field of statistics an essential ally. Yes, it has played a part in bringing science to the brink, but just as surely, it will play a part in bringing science back.

Chapter 19

THE SCOREBOARD WARS

WHEN IS A MEASURE NOT A MEASURE?

During my twenties, I taught perhaps 5000 classroom lessons. No more than 15 were observed by other adults. Even in our age of "school accountability," the classroom remains an obscure, dim-lit, poorly explored place. I understand why politicians and outsiders are desperate for a glimpse inside. And I find it fascinating to watch, through their eyes, as they strain to see the inner workings of schools—and to influence them.

1. THE BEST TEACHER IN AMERICA

In 1982, Jay Mathews was the Los Angeles bureau chief for the *Washington Post*. In theory, that meant covering the biggest stories in the western United States.

In practice, it meant crashing high school calculus classes.

Day after day, Mathews couldn't help dropping by the East Los Angeles classroom of teacher Jaime Escalante. The Bolivian immigrant was a force of nature. With insult comedy, tough love, Spanish catchphrases, and relentless expectations, Escalante helped the students of Garfield High School achieve unparalleled results on the Advanced Placement Calculus exam. These pupils weren't rich or advantaged. Of the 109 that Mathews surveyed, only 35 had a parent with a high school diploma. Yet in one of the nation's toughest courses, they thrived. By 1987, this one school accounted for more than a quarter of all Mexican Americans passing AP Calculus nationwide. By 1988, Escalante would be the most famous teacher in the country: name-dropped in a presidential debate by George H. W. Bush; portrayed in *Stand and Deliver* by Edward James Olmos (who got an Oscar nomination for the role); and profiled by Mathews in a book titled *Escalante: The Best Teacher in America*.

Apart from the quotient rule, Mathews learned a clear lesson from Escalante. Students excel when pushed. A good classroom is a challenging one. And so, hoping to compare schools along this axis, Mathews sought a statistic. Which schools—regardless of socioeconomics and demographics—are doing the most to challenge their students?

He rejected "average AP score." This stat, he believed, spotlighted schools that limit APs to the strongest few, shutting out the bulk of typical students. It struck Mathews as extraordinary and perverse that schools would try to block students from an intellectual challenge. He wanted to measure an AP program's reach, not its exclusivity.

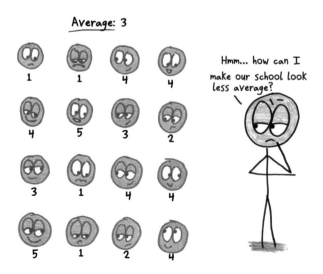

Average: 4.5

Nor did he want to count "average number of AP exams passed." In his understanding, that tended to correlate with socioeconomic status. An AP class was supposed to leave you better prepared for college than when you began, no matter whether you passed. The experience outweighed the score.

In the end, Mathews chose something even simpler: the number of AP (and other college-level) exams taken per graduating student. The scores didn't matter, just the attempt. He dubbed it the Challenge Index. A list of top-performing schools appeared in *Newsweek* in 1998, and again in 2000, and once again in 2003, this time as the featured cover story.

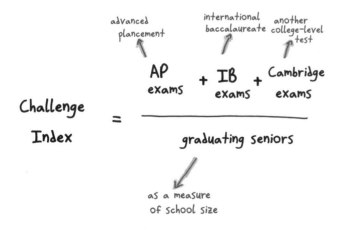

From the start, the rankings courted controversy. One *Newsweek* reader called them "sheer mockery." A professor of education called the list "a disservice to the thousands of schools in which devoted teachers bust their chops daily to provide a challenging and appropriate education for millions of young people who, for many valid reasons, will never see the inside of an AP or IB classroom."

It's been 20 years. The list now appears annually in the *Washington Post*, and Mathews stands by his methods. "I rank," he has written, "with the hope that people will argue about the list and in the process think about the issues it raises."

Maybe this makes me a gullible fish, but I'll take the bait. I think the Challenge Index brings up profound questions—not just about our educational priorities but about the struggle to quantify a messy, multifaceted world. Should we use complex measures or simple ones? What's the trade-off between sophistication and transparency? And most of all, are statistics like the Challenge Index attempts to measure the world as it is or to transform it into something new?

2. THE HORROR SHOW OF BAD METRICS

There are two kinds of people in life: those who like crude dualities and those who do not. And now that I've unmasked myself as the first, please allow me to introduce a statistical distinction I find helpful: *windows* vs. *scoreboards*.

A "window" is a number that offers a glimpse of reality. It does not feed into any incentive scheme. It cannot earn plaudits or incur punishments. It is a rough, partial, imperfect thing—yet still useful to the curious observer. Think of a psychologist asking a subject to rate his happiness on a scale from 1 to 10. This figure is just a crude simplification; only the most hopeless "1" would believe that the number *is* happiness.

Or imagine you're a global health researcher. It's not possible to quantify the physical and mental well-being of every human in a country. Instead, you look to summary statistics: life expectancy, childhood poverty, Pop-Tarts per capita. They're not the whole reality, but they're valuable windows into it.

doesn't show everything

Still better than a wall

The second kind of metric is a "scoreboard." It reports a definite, final outcome. It is not a detached observation, but a summary judgment, an incentive scheme, carrying consequences.

Think of the score in a basketball game. Sure, bad teams sometimes beat good ones. But call the score a "flawed metric for team quality," and people will shoot you the side-eye. You don't score points to prove your team's quality; you improve your team's quality to score more points. The scoreboard isn't a rough measure, but the desired result itself.

Or consider a salesperson's total revenue. The higher this number, the better you've done your job. End of story.

Uh-oh, I've got to catch up... it's almost the last fiscal quarter...

A single statistic may serve as window *or* scoreboard, depending on who's looking. As a teacher, I consider test scores to be windows. They gesture at the truth but can never capture the full scope of mathematical skills (flexibility, ingenuity, affection for "sine" puns, etc.). For students, however, tests are scoreboards. They're not a noisy indicator of a nebulous long-term outcome. They *are* the outcome.

Many statistics make valuable windows but dysfunctional scoreboards. For example, heed the tale of British ambulances. In the late 1990s, the UK government instituted a clear metric: the percentage of "immediately life-threatening" calls that paramedics reached within eight minutes. The target: 75%.

Nice window. Awful scoreboard.

First, there was data fudging. Records showed loads of calls answered in seven minutes and 59 seconds; almost none in eight minutes and one second. Worse, it incentivized bizarre behavior. Some crews abandoned their ambulances altogether, riding bicycles through city traffic to meet the eight-minute target. I'd argue that a special-built patient-transporting truck in nine minutes is more useful than a bicycle in eight, but the scoreboard disagreed.

Allow me to drive this theme home with a series I call "The Horror Show of Bad Metrics":

CLICKS

CALL RESPONSE TIMES

We now handle 90% of complaints in 45 seconds or less!

We must be the most efficient call center in the country!

uh-oh, 41 seconds...
OKAY NICE CHATTING TO YOU GOODBYE FOREVER

DEFINITIONS OF POVERTY

As promised, the governor has completely eliminated poverty in the state.

By officially redefining "poverty" as "being a musical orphan with a diet of gruel and a Cockney accent!

Yup. No poverty 'round these parts.

STUDENT GROWTH

Student scores increased dramatically
from the pre-test to the post-test.

Wow! That must have been some
amazing instruction in between!

Pre-Test

Name:_____

1. What is the
nature of God?

END

Post-Test

Name:_____

1. Please count
the letters in
your own name.

END

NUTRITION

C'mon, it's the healthiest
snack imaginable! Zero
trans fat!

SALES REVENUE

Look! Sales revenues are waaaaay up this month!

Isn't that because we painted our engineers silver and sold them into slavery as "luxury domestic androids"?

Well, if they'd built a proper domestic android from the start, that wouldn't have been necessary.

I just wonder if we undercharged...

please help

UNIVERSITY RANKINGS

Congratulations! Yours is objectively the best university in the country!

You haven't even seen our teaching. You're just saying that because a rich alum paid for a froyo machine in every dorm room.

Exactly. We prefer hard measures, like soft serve.

HIRING PRACTICES

Uh... now picture a box around me...

Why did you hire this terrible mime?

I'm as surprised as you are. He nailed the phone interview.

SURVIVAL RATE

Don't worry. This doctor has a 99% survival rate. You'll be in good hands.

Full disclosure: For the last decade I've specialized in stubbed toes. But in that time, I've only lost three patients.

TEACHER'S VALUE ADDED

Sorry, pal. You're just a bad teacher.

But... why? How do you know?

Well... taking each of your students, and computing their score on this year's test as a percentile amongst all students statewide who achieved the same score as they did on <u>last</u> year's test, then averaging across your classes, you got a low score.

But... what does that mean?

Mathematically? Hell if I know. Practically? You're fired.

Turning back to Mathews and *Newsweek*, I want to confront the natural question: what kind of metric is the Challenge Index?

3. WINDOW OR SCOREBOARD?

In his 1998 introduction to the very first index, Mathews wrote:

> *Nearly every professional educator will tell you
> that ranking schools is counterproductive,
> unscientific, hurtful, and wrong. Every likely
> criteria you might use in such an evaluation is
> going to be narrow and distorted . . . I accept all
> those arguments. Yet as a reporter and a parent,
> I think that in some circumstances a ranking
> system, no matter how limited, can be useful.*

The key word here is "limited." Schools are characterized by an irreducible complexity, like ecosystems or daytime soap operas. To capture such complicated workings with a single metric, you've got two basic options: (1) fuse together many variables into a sophisticated composite score; or (2) just pick a single clear, easy-to-understand variable.

I'm reminded, like the American Neanderthal I am, of football. One simple way to measure a quarterback's performance is completion percentage. How many of his passes are successfully caught? Most seasons, the league leader winds up near 70%; the league average is closer to 60%.

Like many windows, completion percentage straddles a line between "simple" and "simplistic." It treats a conservative 5-yard pass on par with a game-changing 50-yarder: both are "caught." It puts the mild disappointment of a dropped pass on par with the calamity of an intercepted one: both are "not caught." Still, if all statistics are flawed, at least these flaws are transparent. You can't accuse "completion percentage" of false advertising.

At the other end of the spectrum is "passer rating," a dizzying Frankenstat that incorporates passing attempts, completions, yards, touchdowns, and interceptions. It runs from a minimum of zero to a maximum of 158⅓. It is tightly correlated with team victory, and I have never met anyone who claimed to understand how it is computed or what its blind spots are.

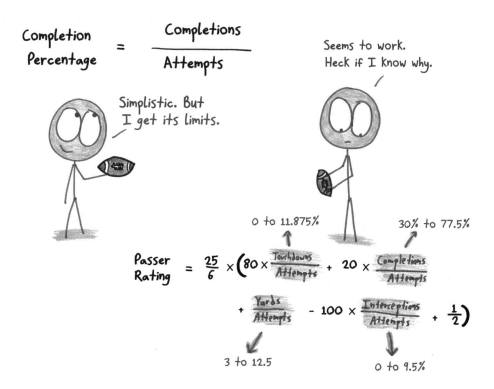

That's the trade-off: sophistication vs. transparency. Passer rating vs. completion percentage. It's clear to me that Mathews is a "completion percentage" kind of guy. In introducing the 2009 *Newsweek* list, he wrote:

> *One of its strengths is the narrowness of the criteria. Everyone can understand the simple arithmetic that produces a school's Challenge Index rating and discuss it intelligently, as opposed to ranked lists like* U.S. News & World Report's *"America's Best Colleges," which has too many factors for me to comprehend.*

All of this casts his Challenge Index as a rough measure: better than nothing, and even better for wearing its imperfections on its sleeve. An honest window.

But when you publish your statistic in a national news magazine under the headline "America's Best High Schools"—well, it starts feeling an awful lot like a scoreboard.

"The list has taken on a life of its own," wrote a National Research Council committee in 2002. "It is now so important to be included among the top 100 schools on the list that some competitive high schools not included have posted disclaimers on their Web sites indicating why this is the case."

"The parents are making the most noise," said a teacher in Milwaukee, Wisconsin. "Your standing in the community will go up because you're offering more AP classes and you might end up on the *Newsweek* top 100."

One hallmark of bad scoreboards is that they're easy to game. In the case of the Challenge Index, you can push students into AP classes. "Because your index only considers the number of AP tests taken, and not the actual scores," wrote Mathews's *Washington Post* colleague Valerie Strauss, "schools put as many kids into the test pipeline as they could."

Let's go, kids! Climb into the AP Tube! Won't this be a fun, college-level adventure?

Another problem lay in the fraction. For convenience's sake, Mathews had used "graduating seniors" as his denominator, rather than "all students." Assuming every student graduates in four years, the math is equivalent. But a high dropout rate yields a perverse reward. If three students each take an AP exam and two later drop out, then by Mathews's math, the remaining graduate has taken three APs.

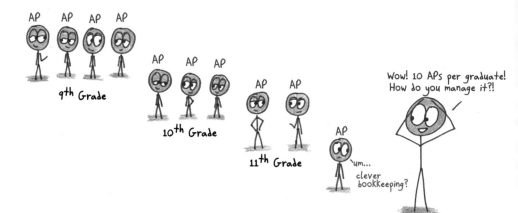

Here, then, is one way to tell the tale of the Challenge Index. It began as a good window. By considering exams taken rather than exams passed, it looked beyond wealth and privilege to the deeper question of whether students were being challenged. Flawless? No. Valuable? Yes.

Then it grew. No longer was it a lone journalist identifying the "most challenging" schools. Now it was a prominent newsmagazine crowning the "best." This created perverse incentives and strange outcomes, turning the good window into a bad scoreboard.

We could tie a bow on our story right there and retire to watch football and/or study for AP exams. But that would miss the most interesting twist—and the real nature of the game that Mathews is playing.

4. UNLEASHING THE TRIBAL PRIMATE

Typically, consumer rankings help to inform specific choices: which car to buy, which university to apply to, which movie to watch. But it's unclear this logic holds for a nationwide ranking of high schools. Am I going to move my family from Florida to Montana in pursuit of a *Newsweek*-approved education? Are you going to consult the stats before deciding between Springfield, Illinois, and Springfield, Massachusetts? Who, exactly, is the index for?

By Mathews's own admission, it's simple. He ranks for ranking's sake.

"We cannot resist looking at ranked lists," he has said. "It doesn't matter what it is—SUVs, ice-cream stores, football teams, fertilizer dispensers. We want to see who is on top and who is not." In 2017, he wrote, "We are all tribal primates endlessly fascinated with pecking orders." The

Challenge Index aims to exploit that quirk of primate psychology, to weaponize it in the cause of making schools more challenging.

Critics call the list easy to game, but Mathews doesn't mind. In fact, that's the whole point: the more students taking the exams, the better. Schools that urge, cajole, and incentivize aren't cheating; they're doing right by their students. He's even happy with the title "best," telling the *New York Times* that the word is "a very elastic term in our society."

As support for his view, Mathews likes to cite a 2002 study of more than 300,000 students in Texas. Focusing on those with low SAT scores, researchers found that students scoring a 2 on an AP exam (a nonpassing grade) later outperformed comparable peers who hadn't attempted the AP at all. The effort itself—even without a passing score—seemed to lay the groundwork for success in college.

All of this flips the narrative on its head. Mathews seems to envision the Challenge Index as a flawed window, as well as the scoreboard the country needs.

For better or worse, the list's impact is real. Mathews has always drawn a line at 1.000—one AP exam per graduating student. In 1998, only 1% of schools nationwide qualified. As of 2017, that number is 12%. In Washington DC, at the epicenter of Mathews's influence (after all, he writes for the *Washington Post*), it is over 70%.

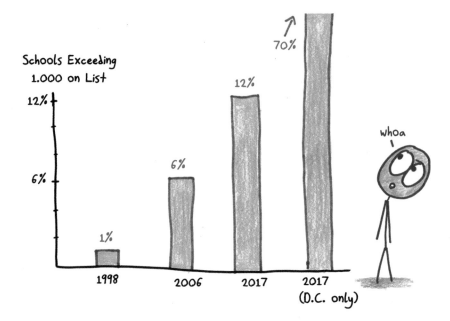

Schools Exceeding
1.000 on List

12%

70%

12%

6%

6%

1%

1998   2006   2017   2017
                       (D.C. only)

whoa

To Mathews, the Challenge Index is a pointed attack on a lethargic and wrongheaded status quo: "the view that schools with lots of rich kids are good, and schools with lots of poor kids are bad." He points with pride to high-ranked schools full of low-income families. He waves off objections—what about the kids at Eastside High in Gainesville, Florida, many of whom read below grade level, or the kids at Locke High in Los Angeles, who drop out at alarming rates?—by saying that these schools deserve recognition for their efforts, not censure for their struggles.

Every statistic encodes a vision of the world it seeks to measure. In the case of the Challenge Index, that vision is tinged with memories of Jaime Escalante and hopes of replicating his approach across the nation. How you feel about Mathews's statistic boils down, in the end, to how you feel about his vision.

Chapter 20

THE BOOK SHREDDERS

There is a beast afoot in the library of life, a chimera known as the Digital Humanities. It has the body of a literary critic, the head of a statistician, and the wild hair of Steven Pinker. Some hail it as a burst of light into a dark cavern. Others scorn it like a slobbering dog with its teeth sunk into a first edition of *Madame Bovary*. What does the creature do?

Simple: it turns books into data sets.

1. WHAT COULD CONCEIVABLY GO WRONG?

Last year, I read Ben Blatt's delightful book *Nabokov's Favorite Word Is Mauve*, which analyzes literature's great authors via statistical techniques. The first chapter, titled "Use Sparingly," explores a cliché of writing advice: avoid adverbs. Stephen King, for example, once compared them to weeds, and warned "the road to hell is paved with adverbs."

So Blatt tallied the rate of *-ly* adverbs ("firmly," "furiously," etc.) in works by various authors. This is what he found:

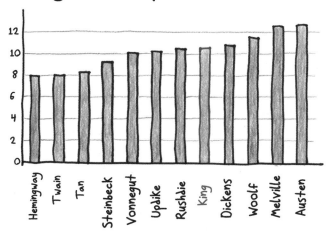

The adverb-friendliness of Austen, one of the language's finest novelists, seemed an adequate refutation of that advice. But then Blatt pointed to a funny pattern. Within a given author's body of work, the greatest novels often used the fewest adverbs.

(See the endnotes for how "greatness" was measured.)

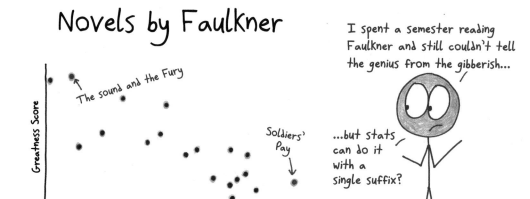

Novels by Faulkner

I spent a semester reading Faulkner and still couldn't tell the genius from the gibberish...

...but stats can do it with a single suffix?

Greatness Score

The sound and the Fury

Soldiers' Pay

Adverbs per 1000 Words

3   5   7   9   11   13   15

F. Scott Fitzgerald's lowest-adverb novel? *The Great Gatsby*. Toni Morrison's? *Beloved*. Charles Dickens's? *A Tale of Two Cities*, followed by *Great Expectations*. Sure, there were exceptions—Nabokov's adverb rate peaks in *Lolita*, arguably his best-regarded novel—but the trend was clear. Low adverb rates yielded clear, powerful writing. High adverb rates suggested flabby second-tier stuff.

I was reminded of one day in college, when my roommate Nilesh smiled and told me, "You know what I love? How often you use the word 'conceivably.' It's one of your signature words."

I froze. I reflected. And in that moment, the word "conceivably" dropped out of my vocabulary.

Language in Its Natural Habitat

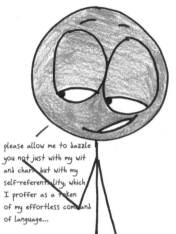

please allow me to dazzle you not just with my wit and charm but with my self-referentiality, which I proffer as a token of my effortless command of language...

VS.

Language under the Statistical Microscope

I – uh – the – I mean – well...

Nilesh mourned the loss for months while I wrestled with the guilt of having betrayed two friends at once, both the word and the roommate. I couldn't help it. The ghost in my brain, the one that turns meanings into words, works by instinct and thrives in shadow. Calling attention to a particular word choice spooked the ghost. It retreated.

Presented with Blatt's statistics, it happened again. I became adverb-paranoid. Ever since, I've been writing like a restless fugitive, afraid those -*ly*s may slip into my prose like spiders crawling into my sleeping mouth. I recognize this is a stilted, artificial approach to language, not to mention a naïve correlation-equals-causation approach to statistics. But I can't help it. It's the promise and the danger of the digital humanities all in a nutshell—with the emphasis, in my case, on "nut."

Taken as a collection of words, literature is a data set of extraordinary richness. Then again, taken as a collection of words, literature is no longer literature.

Statistics work by eliminating context. Their search for insight begins with the annihilation of meaning. As a lover of stats, I'm drawn in. As a lover of books, I flinch. Is there peace to be made between the rich contextuality of literature and the cold analytical power of stats? Or are they, as I sometimes fear, born enemies?

2. STATISTICIANS WILL BE GREETED AS LIBERATORS

In 2010, 14 scientists (led by Jean-Baptiste Michel and Erez Lieberman Aiden) published a blockbuster research article titled "Quantitative Analysis of Culture Using Millions of Digitized Books." Whenever I read its opening line, I can't help whispering, *Daaaaaamn.* It begins, "We constructed a corpus of digitized texts containing about 4% of all books ever printed."

Daaaaaamn.

Like all statistical projects, this required aggressive simplification. The authors' first step was to decompose the entire data set—5 million books, totaling 500 billion words—into what they called "1-grams." They explained: "A 1-gram is a string of characters uninterrupted by a space; this includes words ('banana,' 'SCUBA') but also numbers ('3.14159') and typos ('excesss')."

Sentences, paragraphs, thesis statements—they all vanish. Only snippets of text remain.

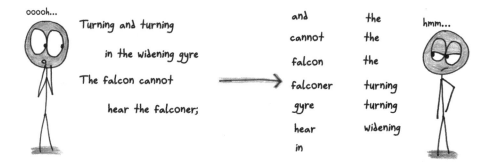

To probe the depths of their data, the authors compiled a list of every 1-gram with a frequency of at least one per billion. Looking at the start, middle, and end of the 20th century, their corpus revealed a growing language:

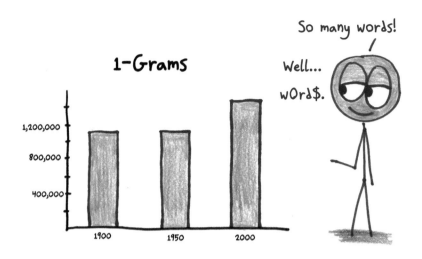

Upon inspection, less than half of the 1-grams from the year 1900 turned out to be actual words (as opposed to numbers, typos, abbreviations, and the like), while more than 2/3 of the 1-grams from the year 2000 were words. Extrapolating from hand-counted samples, the authors estimated the total number of English words in each year:

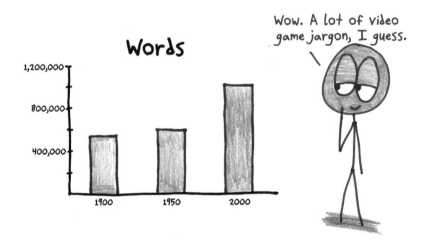

Then, cross-referencing these 1-grams with two popular dictionaries, they found that lexicographers are struggling to keep pace with the language's growth. In particular, the dictionaries miss most of the rarer 1-grams:

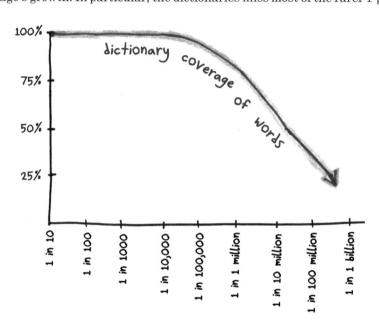

In my own reading, I don't run into many of those rare nondictionary words. That's because . . . well . . . such words are rare. But the language is populated by vast, silent ranks of one-in-a-hundred-million obscurities. Overall, the authors estimated that "52% of the English lexicon—the majority of the words used in English books—consists of lexical 'dark matter' undocumented in standard references." The dictionaries only scratch the surface, missing out on gems like "slenthem."

For these researchers, spelunking through the lexicon was a mere warm-up. The authors went on to examine the evolution of grammar, the trajectories of fame, the fingerprints of censorship, and the shifting patterns of historical memory. All this they accomplished in just a dozen pages, mainly by tracking the frequencies of well-chosen 1-grams.

The paper dropped jaws beyond mine. The journal *Science*, sensing its magnitude, made it free to nonsubscribers. "A New Window on Culture," declared the *New York Times*.

Literary scholars tend to study an exclusive "canon," an elite handful of authors who enjoy deep, focused analysis. Morrison. Joyce. The cat who sat on Joyce's keyboard and produced *Finnegans Wake*. But this paper gestured toward another model: an inclusive "corpus," in which a far-flung population of books, from the celebrated to the obscure, can all share attention. Statistics could serve to topple the literary oligarchy and to instate democracy.

Now, there's no reason both approaches can't live side by side. Close reading *and* statistics. Canon *and* corpus. Still, phrases like "precise measurement" point to a conflict. Can the meanings of literature be made "precise"? Can they be rendered "measurable"? Or will these powerful new tools lead us away from the hard-to-quantify depths of art, in search of nails that our hammer can strike?

3. THIS SENTENCE IS WRITTEN BY A WOMAN

I tend to think of prose as an androgynous thing. Mine is androgynous like a sea sponge; Virginia Woolf's, like a galaxy or a divine revelation. But in Woolf's *A Room of One's Own*, she takes the opposite view. By 1800, she asserts, the prevailing literary style had evolved to house a man's thoughts, rather than a woman's. There was something gendered in the pace and pattern of the prose itself.

That idea rattled around my skull for a few months, until I came across an online project called Apply Magic Sauce. Among other algorithmic feats, it could read copy-and-pasted excerpts from your writing, and, via mysterious analysis, predict your gender.

I had to try it.

From My Blog	Score	Emotional Reaction	
"About" page	90% feminine	Strangely vindicated	oh yeah.
Most popular post (Ultimate Tic-Tac-Toe)	96% masculine	Guilty, somehow	
First-ever viral post (What It Feels Like to Be Bad at Math)	50% masculine, 50% feminine	Too cool for school	

In an internet daze, I spent an hour copying and pasting 25 blog posts, written from 2013 to 2015. The final results looked like this:

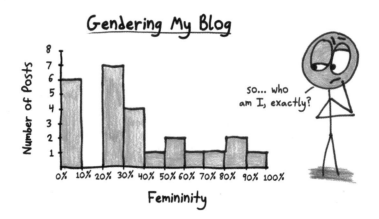

Since the Apply Magic Sauce team guards the secret of its technique, I began snooping around to figure out how the algorithm might operate. Did it diagram my sentences? Sniff out the latent patriarchy of my sentiments? Did it infiltrate my thoughts, the way I imagined Virginia Woolf could have, elevating book reading into a form of soul reading?

Nope. Chances are that it just looked at word frequencies.

In a 2001 paper titled "Automatically Categorizing Written Texts by Author Gender," three researchers managed to achieve 80% accuracy in distinguishing male from female writers, just by counting the occurrences of a handful of simple words. A later paper, titled "Gender, Genre, and Writing Style in Formal Written Texts," laid out these differences in plain terms. First, men use more noun determiners ("an," "the," "some," "most" . . .). Second, women use more pronouns ("me," "himself," "our," "they" . . .).

Word Types in Nonfiction

	Pronouns	Noun Determiners
Men	2.8%	12.5%
Women	3.9%	11.5%

~what a surprise!

oh my word!

Even the frequency of the single innocuous word "you" gives a clue about the author's gender:

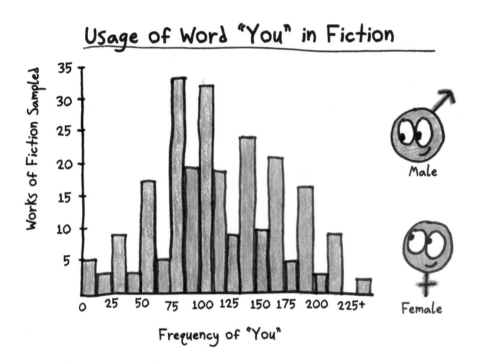

Usage of Word "You" in Fiction

The system's accuracy is made more impressive by its utter simplicity. The approach disregards all context, all meaning, to focus on a tiny sliver of word choices. As Blatt points out, it would evaluate the sentence "This sentence is written by a woman" as most likely written by a man.

If you expand your sights to include all words, and not just little grammatical connectors, then the results take a turn toward the stereotypical. When a data company called CrowdFlower trained an algorithm to infer the gender of the owners of Twitter accounts, it spat out the following list of gender-predictive words:

Most Male

Wrestling

#thisiswhyweplay

director

producing

blushing

Martin

notes

Celtic

Patrick

dad

Most Female

♡

camgirl

makeup

hurry

actress

communication

I

mommy

yours

psych

And in *Nabokov's Favorite Word Is Mauve*, Ben Blatt finds the most gender-indicative words in classic literature:

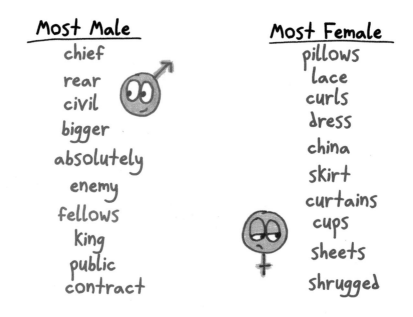

Most Male

chief

rear

civil

bigger

absolutely

enemy

fellows

king

public

contract

Most Female

pillows

lace

curls

dress

china

skirt

curtains

cups

sheets

shrugged

It seems that Apply Magic Sauce relies on these sorts of clues, too. When mathematician Cathy O'Neil tried the algorithm on a man's writing about fashion, he scored 99% female. When she checked it on a woman's writing about mathematics, she scored 99% male. Three of O'Neil's own pieces scored 99%, 94%, and 99% male. "That's not an enormous amount of testing," she wrote, "but I'm willing to wager that this model represents a stereotype, assigning the gender of the writer based on the subject they've chosen."

The inaccuracies don't quiet my creepy feeling. It seems my maleness so pervades my thinking that an algorithm can detect it in two nonoverlapping ways: my pronoun usage and my fondness for Euclid.

I know that, on some level, this vindicates Woolf. She saw that men and women experience different worlds, and believed that the struggle to give women a voice had to begin all the way down at the level of the sentence. Crude statistics bear this out: women write about different topics, and in different ways, than men do.

Still, I find it all a little depressing. If Woolf's writing reveals her femaleness, I like to think it's embedded in her wisdom and her humor, not in her low density of noun determiners. To hear Woolf distinguish masculine from feminine prose feels like going to a trusted doctor. To have an algorithm do the same feels like getting a pat-down at the airport.

4. THE BUILDING, THE BRICKS, AND THE MORTAR

The Federalist Papers, written in 1787, helped to define American governance. They're full of political wisdom, shrewd argumentation, and timeless catchphrases ("spectacles of turbulence and contention," anyone?). They'd make a killer line on the CV, except for one problem.

The authors didn't sign their names.

Of the original 77, historians could attribute 43 to Alexander Hamilton, 14 to James Madison, five to John Jay, and three to joint authorship. But 12 remained a mystery. Were they Hamilton's or Madison's? Almost two centuries later, the case had long since gone cold.

Come the 1960s, and enter two statisticians: Frederick Mosteller and David Wallace. Fred and Dave recognized the subtlety of the problem.

Hamilton's sentences averaged 34.55 words; Madison's, 34.59. "For some measures," they wrote, "the authors are practically twins." And so they did what good statisticians do when confronting a vexing problem.

They tore *The Federalist Papers* to shreds.

Context? Gone. Meaning? Annihilated. As long as *The Federalist Papers* remained a collection of foundational texts, they were useless. They had to become strips of paper, a pile of tendencies—in other words, a data set.

Even then, most words proved useless. Their frequency depended not on the author, but on the topic. Take "war." "In discussions of the armed forces the rate is expected to be high," Fred and Dave wrote. "In a discussion of voting, low." They labeled such words "contextual" and sought at all costs to avoid them. They were too meaningful.

Their search for meaninglessness struck gold with "upon," a word that Madison almost never used but that Hamilton treated like an all-purpose seasoning:

Armed with their data, Fred and Dave could reduce each author to something like a deck of cards, dealing out words at predictable rates. Then, by tallying word frequencies in the disputed papers, they could infer from which deck the text had been dealt.

It worked. "The odds are enormously high," they concluded, "that Madison wrote the 12 disputed papers."

In the half century since, their techniques have become standard. They have helped explore the authorship of ancient Greek prose, Elizabethan sonnets, and speeches by Ronald Reagan. Ben Blatt applied the algorithm almost 30,000 times, using just 250 common words, to determine which of two authors had written a given book. His success rate was 99.4%.

My intellect knows there's nothing wrong with this. Still, my emotions rebel. How can you understand a book by shredding it to bits?

In 2011, a team of authors at the Stanford Literary Lab attempted a tricky leap: from fingerprinting *authors* to fingerprinting *genres*. They used two methods: word-frequency analysis and a more sophisticated sentence-level tool (called Docuscope). To their surprise, both methods converged on accurate genre judgments.

Take this excerpt, from a passage the computer identified as the most "Gothic" page in their entire 250-novel corpus:

He passed over loose stones through a sort of court, till he came to the arch-way; here he stopped, for fear returned upon him. Resuming his courage, however, he went on, still endeavouring to follow the way the figure had passed, and suddenly found himself in an enclosed part of the ruin, whose appearance was more wild and desolate than any he had yet seen. Seized with unconquerable apprehension, he was retiring, when the low voice of a distressed person struck his ear. His heart sunk at the sound, his limbs trembled, and he was utterly unable to move. The sound which appeared to be the last groan of a dying person, was repeated . . .

I find this to be two kinds of creepy. First, there's the Gothic creepiness of ruined archways and deathly groans. And second, there's the creepiness of a computer detecting the Gothic air without even glancing at the words "arch-way," "ruin," or "last groan of a dying person." It had tagged the passage on the basis of pronouns ("he," "him," "his"), auxiliaries ("had," "was"), and verb constructions ("struck the," "heard the").

I'm unnerved. What does the algorithm know that I don't?

To my relief, the authors offered a tentative answer. There is no single element that distinguishes a writer or genre, no unique feature from which all others follow. Rather, writing has many distinctive features, running from the galactic structure of the novel down to the molecular structure of the syllable. Statistical tendencies and larger meanings can coexist, living side by side in the same sequence of words.

Most of the time, I read for architecture. Plot, theme, character. It's the high-level structure: the aspects visible to any passerby but impenetrable to statistics.

 =

If I look closer, I can see the brickwork. Clauses, sentence constructions, the design of a paragraph. It's the microlevel structure that my high school English teachers taught me to scrutinize. Computers can learn to do the same.

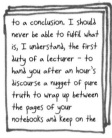 =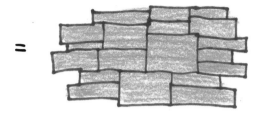

to a conclusion. I should never be able to fulfil what is, I understand, the first duty of a lecturer – to hand you after an hour's discourse a nugget of pure truth to wrap up between the pages of your notebooks and keep on the

And beneath this, hidden from my view, is the mortar. Pronouns, prepositions, indefinite articles. It's the nanolevel structure, the limestone cement that holds it all together, too subtle for my eyes, but ideal for the chemical analysis of a statistician.

... you after an hour's discourse a... =

I know it's just a metaphor, but metaphor is the language that the ghost in my brain speaks. Heartened, I grabbed the first section of this book ("How to Think Like a Mathematician") and sampled the rate of *-ly* adverbs. It landed at 11 per 1000 words—around the same as Virginia Woolf, which I took as a friendly omen. Then, unable to help myself, I purged the document of extraneous *-ly*s until the rate dipped below eight per 1000. That territory belongs to Ernest Hemingway and Toni Morrison. I was cheating, and it felt great.

Can the new statistical techniques find harmony with the older, richer, more human ways of understanding language? Yes; conceivably.

$(1-\lambda)\ ^{2}\pi\ \sqrt{a^{2}+b^{2}}$ $|\underline{u}\wedge\underline{v}|=|\underline{u}||\underline{v}|\sin\theta$ $|AB|:|BC|=\lambda:(1-\lambda)\ ^{2}\pi\ \sqrt{a^{2}+b^{2}}$ $|\underline{u}\wedge\underline{v}|=|\underline{u}||\underline{v}|\sin\theta$

$\underline{b}=\underline{b}-\underline{a}=\lambda(c-a)$ $(\underline{u}\cdot\underline{v})^{2}+|\underline{u}\wedge\underline{v}|^{2}=|\underline{u}|^{2}|\underline{v}|^{2}$ $\vec{AC}=(\underline{c}-\underline{a}),\ \vec{AB}=\underline{b}-\underline{a}=\lambda(c-a)$ $(\underline{u}\cdot\underline{v})^{2}+|\underline{u}\wedge\underline{v}|^{2}=$

$\underline{c}\quad \lambda=\frac{1}{2}$ $|\underline{u}\wedge\underline{v}|^{2}=|\underline{u}|^{2}|\underline{v}|^{2}\sin^{2}\theta$ $b=(1-\lambda)g+\lambda\underline{c}\quad \lambda=\frac{1}{2}$ $|\underline{u}\wedge\underline{v}|^{2}=|\underline{u}|^{2}|\underline{v}|$

$\underline{u}\wedge\underline{v}=|\underline{u}||\underline{v}|\sin\theta\underline{n}\quad 2\pi\sqrt{a^{2}+b^{2}}\quad \underline{b}=\frac{1}{2}(\underline{a}+\underline{c})$ $\underline{u}\wedge\underline{v}=|\underline{u}||\underline{v}|\sin$

$\underline{u}\wedge\underline{v}=A\underline{n}$
$A=|\underline{u}||\underline{v}|\sin\theta$
$t+\delta t$

$\underline{d}=\frac{1}{2}(\underline{b}+\underline{c}),\ \underline{g}=\frac{1}{3}(\underline{a}+\underline{b}+\underline{c})$

$);\ \underline{g}=\frac{1}{3}(\underline{a}+\underline{b}+\underline{c})$

area A
$|\underline{u}||\underline{v}|\sin\theta$

$\delta\underline{r}=\underline{r}(t+\delta t)-\underline{r}(t)$

$dt\to 0$

$\frac{d\underline{r}}{dt}=\lim_{\delta t\to 0}\frac{\delta\underline{r}}{\delta t}$

$(\underline{u}\cdot\underline{v})^{2}+|\underline{u}\wedge\underline{v}|^{2}=|\underline{u}|^{2}|\underline{v}|^{2}$

$|\underline{v}|^{2}=|\underline{u}|^{2}|\underline{v}|^{2}$
$(u_{1}u_{1}+u_{2}u_{2}+u_{3}u_{3})^{2}+(u_{2}u_{3}-u_{3}u_{2})^{2}-$
$(u_{1}v_{2}-u_{2}v_{1})^{2}=$
$(u_{3})^{2}+(u_{2}u_{3}-u_{3}u_{2})^{2}-$
$(u_{1}v_{2}-u_{2}v_{1})^{2}=$
$(u_{1}v_{1}-u_{1}v_{3})^{2}+(u_{1}v_{2}-u_{2}v_{1})^{2}=$
$\frac{2}{3}(U_{1}^{2}+U_{2}^{2}+\frac{2}{3})$
$=(u_{1}^{2}+u_{2}^{2}+u_{3}^{2})(U_{1}^{2}+U_{2}^{2}+\frac{2}{3})$

$(\theta,\varphi)\in[0,2\pi]\times$
$[-\pi/2,\pi/2]$

$x^{2}+y^{2}=z^{2}$

$x^{2}+y^{2}=z^{2}$

$(1-\lambda)\ ^{2}\pi\ \sqrt{a^{2}+b^{2}}$ $|\underline{u}\wedge\underline{v}|=|\underline{u}||\underline{v}|\sin\theta$ $|AB|:|BC|=\lambda:(1-\lambda)\ ^{2}\pi\ \sqrt{a^{2}+b^{2}}$ $|\underline{u}\wedge\underline{v}|=|\underline{u}||\underline{v}|\sin\theta$

$\underline{b}=\underline{b}-\underline{a}=\lambda(c-a)$ $(\underline{u}\cdot\underline{v})^{2}+|\underline{u}\wedge\underline{v}|^{2}=|\underline{u}|^{2}|\underline{v}|^{2}$ $\vec{AC}=(\underline{c}-\underline{a}),\ \vec{AB}=\underline{b}-\underline{a}=\lambda(c-a)$ $(\underline{u}\cdot\underline{v})^{2}+|\underline{u}\wedge\underline{v}|^{2}=$

$\underline{c}\quad \lambda=\frac{1}{2}$ $|\underline{u}\wedge\underline{v}|^{2}=|\underline{u}|^{2}|\underline{v}|^{2}\sin^{2}\theta$ $b=(1-\lambda)g+\lambda\underline{c}\quad \lambda=\frac{1}{2}$ $|\underline{u}\wedge\underline{v}|^{2}=|\underline{u}|^{2}|\underline{v}|$

$\underline{u}\wedge\underline{v}=|\underline{u}||\underline{v}|\sin\theta\underline{n}\quad 2\pi\sqrt{a^{2}+b^{2}}\quad \underline{b}=\frac{1}{2}(\underline{a}+\underline{c})$ $\underline{u}\wedge\underline{v}=|\underline{u}||\underline{v}|\sin$

$\underline{u}\wedge\underline{v}=A\underline{n}$
$A=|\underline{u}||\underline{v}|\sin\theta$
$t+\delta t$

$\underline{d}=\frac{1}{2}(\underline{b}+\underline{c}),\ \underline{g}=\frac{1}{3}(\underline{a}+\underline{b}+\underline{c})$

$);\ \underline{g}=\frac{1}{3}(\underline{a}+\underline{b}+\underline{c})$

area A
$|\underline{u}||\underline{v}|\sin\theta$

$\delta\underline{r}=\underline{r}(t+\delta t)-\underline{r}(t)$

$dt\to 0$

$\frac{d\underline{r}}{dt}=\lim_{\delta t\to 0}\frac{\delta\underline{r}}{\delta t}$

$(\underline{u}\cdot\underline{v})^{2}+|\underline{u}\wedge\underline{v}|^{2}=|\underline{u}|^{2}|\underline{v}|^{2}$

$|\underline{v}|^{2}=|\underline{u}|^{2}|\underline{v}|^{2}$
$(u_{1}u_{1}+u_{2}u_{2}+u_{3}u_{3})^{2}+(u_{2}u_{3}-u_{3}u_{2})^{2}-$
$(u_{1}v_{2}-u_{2}v_{1})^{2}=$
$(u_{3})^{2}+(u_{2}u_{3}-u_{3}u_{2})^{2}-$
$(u_{1}v_{2}-u_{2}v_{1})^{2}=$
$\frac{2}{3}(U_{1}^{2}+U_{2}^{2}+\frac{2}{3})$
$=(u_{1}^{2}+u_{2}^{2}+u_{3}^{2})(U_{1}^{2}+U_{2}^{2}+\frac{2}{3})$

$(\theta,\varphi)\in[0,2\pi]\times$
$[-\pi/2,\pi/2]$

$x^{2}+y^{2}=z^{2}$

$x^{2}+y^{2}=z^{2}$

$(1-\lambda)\ ^{2}\pi\ \sqrt{a^{2}+b^{2}}$ $|\underline{u}\wedge\underline{v}|=|\underline{u}||\underline{v}|\sin\theta$ $|AB|:|BC|=\lambda:(1-\lambda)\ ^{2}\pi\ \sqrt{a^{2}+b^{2}}$ $|\underline{u}\wedge\underline{v}|=|\underline{u}||\underline{v}|\sin\theta$

$\underline{b}=\underline{b}-\underline{a}=\lambda(c-a)$ $(\underline{u}\cdot\underline{v})^{2}+|\underline{u}\wedge\underline{v}|^{2}=|\underline{u}|^{2}|\underline{v}|^{2}$ $\vec{AC}=(\underline{c}-\underline{a}),\ \vec{AB}=\underline{b}-\underline{a}=\lambda(c-a)$ $(\underline{u}\cdot\underline{v})^{2}+|\underline{u}\wedge\underline{v}|^{2}=$

V

ON THE CUSP
THE POWER OF A STEP

T he Fitbit-clad among us have a pretty good idea of how many steps they take: 3000 on a sedentary day; 12,000 on an active one; 40,000 if they're fleeing all day from a very slow bear. (Maybe just four or five if they're fleeing from a fast one.)

This kind of counting masks a truth we all know: Not all steps are created equal.

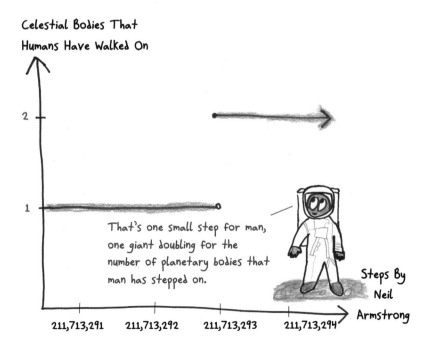

Celestial Bodies That Humans Have Walked On

That's one small step for man, one giant doubling for the number of planetary bodies that man has stepped on.

Steps By Neil Armstrong

211,713,291 211,713,292 211,713,293 211,713,294

Times that Humans Have Harnessed Electricity

1

"Many of life's failures are people who did not realize how close to success they were when they gave up."

Thomas Edison

0

2,893   2,894   2,895   2,896   2,897   2,898

Attempted Light Bulb Designs

camel's back pain

straws

Mathematicians distinguish between two kinds of variables. *Continuous* variables can change by any increment, no matter how small. I can drink 1 liter of diet soda, or 2 liters, or any tooth-dissolving quantity in between. A skyscraper can loom 300 meters, or 300.1, or 300.0298517. Between any two values, no matter how close together, you can always split the difference.

Discrete variables, by contrast, move by jumps. You can have one sibling, or two, but not 1¼. When you buy a pencil, the store can charge 50 cents, or 51 cents, but not 50.43871 cents. With discrete variables, some differences cannot be split.

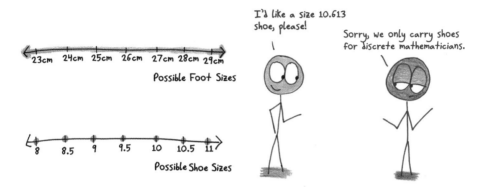

I'd like a size 10.613 shoe, please!

Sorry, we only carry shoes for discrete mathematicians.

23cm 24cm 25cm 26cm 27cm 28cm 29cm

Possible Foot Sizes

8   8.5   9   9.5   10   10.5   11

Possible Shoe Sizes

Life is a funny mixture of continuous and discrete variables. Ice cream is a continuous quantity (giving continual pleasure), and yet it is served in discrete sizes. The quality of job interviews varies continuously, and yet any application yields a discrete number of job offers (zero or one). The speed of traffic is continuous; the speed limit is a single discrete barrier.

This conversion process can amplify tiny increments into huge changes. A tiny burst of acceleration earns you a speeding ticket. A misplaced hiccup during a job interview costs you the offer. A desire for just a *little* more ice cream forces you, through no fault of your own, to order an 18-scoop bucket called "the BellyBuster."

That's what happens in a world that turns "continuous" into "discrete." For every cusp in life, there is a critical turning point—an infinitesimal step with the power to change everything.

Chapter 21

THE FINAL SPECK OF DIAMOND DUST

About 250 years ago, the economist Adam Smith asked a question you might expect from a toddler: why do diamonds cost so much more than water?

Life-Giving Liquid

$ \dfrac{1}{10,000}$

Shiny Rock

$ 10,000

He devoted 100 words to posing the puzzle, followed by 13,000 to developing a theory of prices, by the end of which he still didn't have a solution. Put yourself in the shoes of an alien visitor, and you too may find yourself stumped. What foolish planet would place a lower value on droplets of life-giving H_2O than on hard bits of decorative carbon? Do usefulness and price have no connection whatsoever? Are humans just irredeemably illogical?

The pursuit of these questions transformed economics. Scholars began the journey in the holistic spirit of moral philosophers and ended it with the ruthless rigor of mathematicians. But they did, at last, find an answer. You want to know why we put a higher price on shiny rocks than on the juice that makes our kidneys function?

Easy. Just think on the margin.

1. THE TIME HAS COME, LÉON WALRAS SAID, TO TALK OF MANY THINGS

Classical economics lasted a century, from the 1770s to the 1870s. During that century, brilliant minds helped to push forward the boundaries of human understanding. But I do not come to eulogize classical economics; I come to poke fun at it. Classical economists bit down hard on an idea that, to the modern palate, tastes like dust and mistakes: the "labor theory of value."

The theory states that the price of a good is determined by the labor required to make it.

To tease out this premise, let's travel to a hunter-gatherer society. Here, it takes six hours to hunt a deer, and three hours to gather a basket of berries. According to the labor theory of value, the costs will depend on a single factor: not scarcity, tastiness, or the current diet fads, but the ratio of required labor. A deer takes twice as long to obtain as a berry basket, so it will cost twice as much.

I feel like they're taking me for granted.

You've got to play hard to get.

Of course, labor itself is not the only input. What if deer hunting requires fancy deer-spears (which take four hours to make) while berry gathering requires only a generic basket (one hour to make)? Then we total it up: in all, a deer involves 6 + 4 = 10 hours of labor, and a berry basket only 3 + 1 = 4 hours of labor, so a deer should trade for 2.5 times as much as the berries. Similar adjustments can account for all inputs, even worker training.

Under this theory, everything is labor, and labor is everything.

In this vision of the economy, the "supply side" sets the price, and the "demand side" determines the quantity sold. The logic feels quite natural. When I go to the market to buy deer meat and iPads, I don't get to pick the prices. The seller does that. My only choice is whether or not to buy.

Appealing. Intuitive. And, according to today's best experts, quite wrong.

In the 1870s, economics experienced an intellectual growth spurt. A loose karass of thinkers across Europe, mulling over the successes and struggles of their predecessors, came to feel that vague philosophizing was not enough. They sought to put economics on sturdier foundations. Individual psychology. Careful empiricism. And most of all, rigorous mathematics.

These new economists wrestled with the question of discrete vs. continuous. In reality, I can buy one diamond, or two, but nothing in between. We purchase in discrete chunks.

That's unfortunate, because it's much harder, mathematically, to manipulate quantities that change by leaps and spasms than ones that vary by smooth, continuous growth. Thus, to ease the path, these new economists assumed that we could buy any quantity of a product, down to an infinitesimal increment. Not a whole diamond, but a speck of diamond dust. It's a simplification, of course: not true, but powerfully useful.

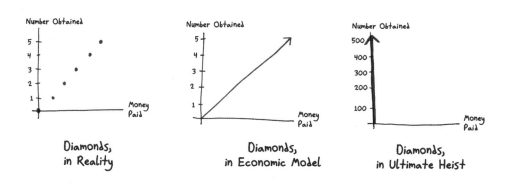

This unlocked a new kind of analysis: Economists began to think on the margin. The question became not "What's the *average* berry basket worth?" but "What is *one extra* berry basket worth?" or better yet "What is the value of a single extra berry?" It was the dawn of modern economics, a moment now known as the "marginal revolution."

Its leaders included William Stanley Jevons, Carl Menger, and an unlikely hero named Léon Walras, later dubbed by a commentator "greatest of all economists." Before achieving that title, he built an eclectic résumé: engineering student, journalist, railway clerk, bank manager, and Romantic novelist. Then, one summer night in 1858, he went for a fateful stroll with his father. Old Papa Walras must have been a persuasive man, because by the end of the evening, Léon had overcome his wanderlust and set his heart on economics.

Classical economists had tackled big, bold questions about the nature of markets and society. Marginalists narrowed the focus to individuals making tiny decisions on the margin. Walras aimed to unify the two levels of analysis, to build a sweeping vision of the entire economy on a foundation of tiny mathematical steps.

I shall begin with simple math,
as economics should.
Two agents trading quantities
of two generic goods.
Then step by step,
the theory grows,
like man from childhood...

2. OF MUFFINS, FARMS, AND COFFEE SHOPS; OF GENTLY LOADED SPRINGS

It's role-play time. Congratulations: you're a farmer!

But I'm afraid you'll find, as farmers do, that the more land you cultivate, the less fertile each additional acre becomes.

Land is heterogeneous. When you begin to farm, you first pick the richest, ripest patches of soil. As you exhaust the best opportunities, each new acre offers slightly lower returns than the acre before. Eventually, only stretches of barren rock remain.

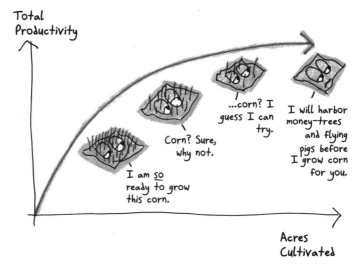

This idea long predates marginalism. One preclassical economist gave a mechanical analogy:

> *The earth's fertility resembles a spring that is being pressed downwards by the addition of successive weights . . . After yielding a certain amount . . . weights that formerly would have caused a depression of an inch or more will now scarcely move it by a hair's breadth.*

The marginalist breakthrough came in expanding the same notion from agriculture to human psychology. For example, just watch me eating cornbread muffins:

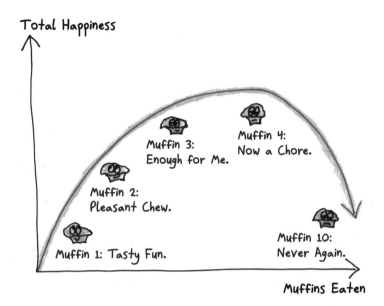

Perhaps all muffins are created equal, but that's not what it feels like as I eat them. The more bites I've consumed, the less delightful each additional bite becomes. Before long, they stop satiating and begin nauseating.

It's not just food. The same principle holds for all consumer goods: scarves, SUVs, even Kurt Vonnegut novels. In each case, it's hard to draw the same marginal pleasure from your 10th item as you did from your first; the benefit of a step depends on how many steps you've already taken. Economists today dub this "the law of diminishing marginal utility," although I prefer to call it "the reason I didn't really enjoy *Breakfast of Champions*."

How fast is the decline? Well, it varies. As leading marginalist William Stanley Jevons wrote:

> *This function of utility is peculiar to each kind of object, and . . . to each individual. Thus, the appetite for dry bread is much more rapidly satisfied than that for wine, for clothes, for handsome furniture, for works of art, or, finally, for money. And every one has his own peculiar tastes in which he is nearly insatiable.*

In addition to revealing a lot about William Stanley Jevons, this passage points toward a new theory of economic demand. People make decisions on the basis of marginal utility.

Imagine a glorious economy with only two goods: muffins and coffee. How should I allocate my spending? Well, with each dollar, I ask the same question: will I draw more joy from adding this to the muffin budget, or the coffee budget? Even a muffin-loving fool like me will eventually prefer a first coffee to an eleventh pastry, and even a caffeine-addicted economist will eventually prefer a first muffin to an eleventh coffee.

This logic boils down to a simple criterion: *In a perfect budget, the last dollar spent on each good will yield exactly the same benefit.*

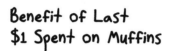
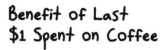

Benefit of Last $1 Spent on Muffins = Benefit of Last $1 Spent on Coffee

You look delicious.

Aww shucks. Well, you look *equally* delicious.

This Walrasian insight marked a new, more psychological foundation for economics. As Jevons wrote, "A true theory of economy can only be attained by going back to the great springs of human action—the feelings of pleasure and pain." For the first time, scholars understood "the economy" to include not just the visible ledger of transactions, but the invisible psychology of preference and desire.

Going further, marginalists judged that sellers function in a similar way to consumers. For example, say you're now a coffee shop owner. How many workers should you hire?

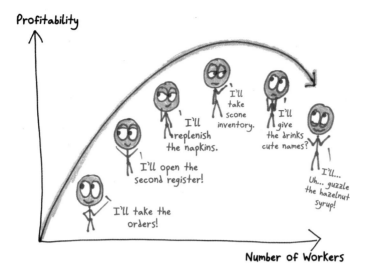

As your business grows, you draw less and less benefit from each new employee. Eventually, you hit a point where adding another worker will cost more in wages than it will bring in coffee sales. At that point, stop hiring.

This logic helps balance the various inputs of a coffee shop. It's pointless to buy a dozen espresso machines without hiring enough workers to operate them, or to pour money into advertising without stocking enough coffee beans to serve the new customers. How can our budget achieve a sensible harmony among these inputs? Simple: Keep buying until your last $1 spent on each input yields $1 of extra productivity. Spend any less, and you'd miss out on possible benefit. Spend any more, and you'd eat into profit.

In school, I learned that economics has mirror symmetry between consumers and producers. Consumers buy goods so as to maximize utility. Producers buy production inputs so as to maximize profits. Each keeps buying until the marginal benefit of the next item would no longer justify the cost. For this tidy parallel framework, we can thank (or, depending how you feel about capitalism, blame) the marginalists.

3. AND WHY A DIAMOND COSTS SO MUCH BUT WATER, NOT A THING

The old labor theory of prices held half a truth. After all, it *does* take more labor to extract diamonds from mines than water from wells. Still, it leaves a key question unanswered: why would anyone exchange huge amounts of money for tiny amounts of carbon?

Well, imagine you're rich. (Perhaps you've played this game before.) You've already got all the water you need. Enough to drink, shower, water the petunias, and maintain the backyard water park. Another $1000 of water? Worthless.

But $1000 of diamond? Novel! Shiny! It's like your first cup of coffee after 11 muffins.

What's true for one rich person is even truer in the aggregate. The first diamond available would find an eager buyer at an absurd price (say, $100,000). The second diamond would find a slightly less eager buyer (for $99,500). As you exhaust the most gung-ho buyers, you've got to lower the price. The more diamonds on the market, the lower the utility of the final diamond.

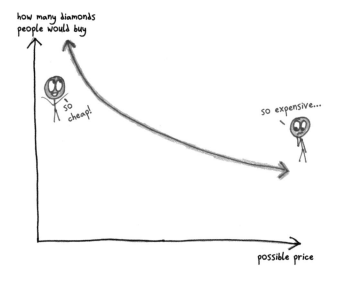

On the supply side, a parallel principle operates. If the first diamond can be mined for $500, the next will cost $502, and so on, each additional diamond costing a bit more to obtain.

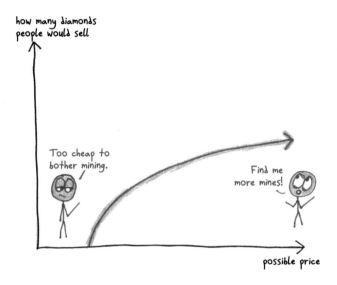

As the market grows, the numbers converge. Inch by inch, the cost of supply increases. Inch by inch, the utility of consumption drops. Eventually, they meet at a single price known as "market equilibrium."

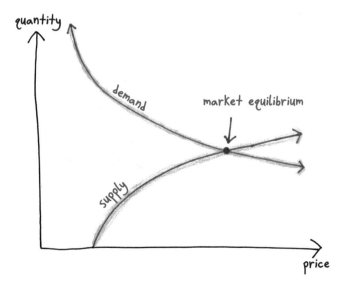

Yes, the economy's first cup of water is worth far more than its first shard of diamond. But prices don't depend on the first increment, or even the average increment. They depend on the *last* increment, on the final speck of diamond dust.

It's a beautiful theory. And, like a too-good-to-be-true factoid or a perfect-skinned model on a magazine cover, it raises a question: Is it real? Do producers and consumers actually think on the margin?

Well, no. To quote Yoram Bauman, an economics PhD and self-described stand-up economist: "Nobody goes to a grocery store and says, 'I'm going to buy an orange. I'm going to buy another orange. I'm going to buy another orange . . . '"

Marginalism doesn't capture people's conscious thought processes any more than the theory of evolution describes the motivations of early mammals. It is an economic abstraction, throwing out the details of reality to provide a useful, simplified map. The measure of a theory is its predictive power—and here, marginalism triumphs. It's not always conscious, but on some abstract level, we are all creatures of the margin.

4. THE MARGINAL REVOLUTION

The marginal revolution marked a point of no return for economics, a historical cusp. It's the moment that econ went mathy.

John Stuart Mill, the last great classical economist, had gestured in this direction. "The proper mathematical analogy is that of an equation," he wrote. "Demand and supply . . . will be made equal." Take the diamond market. If demand exceeds supply, then buyers will compete, driving the price up. If supply exceeds demand, then sellers will compete, driving the price down. In Mill's succinct phrasing: "competition equalizes."

Compelling as Mill's analysis is, the marginalists dissed it as insufficiently mathematical. "If economics is to be a real science at all," Jevons wrote, "it must not deal merely with analogies; it must reason by real equations." Or as Walras says: "Why should we persist in using everyday language to explain things in the most cumbrous and incorrect way . . . as John Stuart Mill does repeatedly . . . when these same things can be stated far more succinctly, precisely, and clearly in the language of mathematics?"

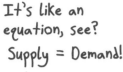

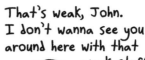

Walras walked the walk. His signature work, *Elements of Pure Economics*, is a mathematical tour de force. It builds from clear assumptions to a sweeping theory of equilibrium. As if to prove his detachment from reality, he devotes more than 60 pages to the analysis of the simplest economy imaginable: two people exchanging fixed quantities of two goods. This work, with its unprecedented rigor and profound abstraction, later earned praise as "the only work by an economist that will stand comparison with the achievements of theoretical physics." Walras would have relished the compliment. His driving goal, ever since the stroll with his father that one evening, was to elevate economics to the level of a rigorous science.

More than a century later, it's worth checking in. How has the mathematical pivot been working out for Walras and his clan?

For better or worse, the marginal revolution science-ified economics in one way. It made it less accessible. Adam Smith and the classicists wrote for a general (if educated) audience. Walras wrote for mathematically skilled specialists. His vision has pretty much won out: PhD programs in economics today prefer a student with a degree in math and a little knowledge in economics to a student with a degree in econ but little mathematical training.

The marginalist revolution can claim another scientific legacy in economics: renewed empiricism. Researchers today agree that economic ideas can't just appeal to intuition or logic; they've got to match real-world observations.

Economics still isn't physics, of course. Human-made systems like markets don't obey tidy mathematical laws. At best, they resemble the swirling complexity of phenomena like the weather and fluid turbulence—precisely the systems that math still struggles to grasp.

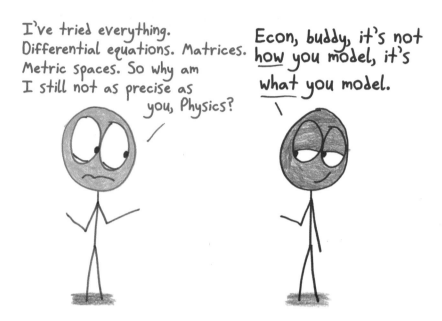

As marginalism edged out competing systems of thought, it brought yet another change: Economics, like its brethren in the natural sciences, became ahistorical. Before marginalism, ideas rose, fell, and recirculated across the decades. You'd read past thinkers in order to soak in their holistic vision of the economy, absorbing their worldview through their language. Today, economists pretty much agree on the foundations and don't care to read ideas in their original, archaic formulations. Give them the streamlined modern version, thank you very much. The marginalists helped economists to stop caring as much about historical thinkers—including, in a final irony, the marginalists.

Chapter 22

BRACKETOLOGY

Sometimes you can reveal a misconception just by uttering a phrase. "Soar like a penguin." Nope, they don't fly. "Famous historical Belgians." Sorry, Belgium's most celebrated native is the waffle. "Too full for dessert." C'mon, nothing justifies refusing cupcakes. However, my favorite error-in-a-single-breath is one you've probably heard countless times:

"Bumped into the next tax bracket."

Those six words encapsulate a genuine, widespread, and wholly misguided fear: What if earning a little extra lifts me into a higher tax bracket? Could I wind up losing money overall?

No, please, no extra shifts! My family is strapped for cash right now.

Um... what?

Why so glum? Bad day at work?

Terrible. I got a raise.

OH NO

I'd like to request a pay cut.

Don't you mean an increase?

Nah, I'm trying to move down a tax bracket.

In this chapter, I'll explain the elementary mathematics behind the income tax, narrate a brief history of its role in American civic life, and identify which Disney characters are its most zealous advocates. But first, I want to reassure you that tax brackets do not go bump in the night. The dreaded "tax bracket bump" is as fictional as a Belgian celebrity.

Our tale begins in 1861. As the American Civil War looms, the federal government begins cooking up get-rich-quick schemes, but none of them hold much promise. The longtime tariff on foreign goods has stopped bringing adequate funds. A tax on consumer purchases might alienate voters by hitting poor Americans harder. And a tax targeting rich people's wealth (including property, investments, and savings) would violate the Constitution's ban on "direct" taxes. What's a cash-strapped Congress to do?

The only thing they could. In August, the federal government introduced a temporary emergency tax on income. On any earnings above $800, you now owed 3%.

The rationale? Money has a diminishing marginal utility. The more dollars you've got, the less another dollar means. Thus, it causes less pain to tax somebody's thousandth dollar than to tax somebody else's first. The resulting systems, wherein higher earnings face a higher marginal tax rate, are called "progressive." (A "regressive" tax draws a larger percentage from those with lower incomes; a "flat" tax takes the same percentage from everybody.)

You can imagine our American forefathers, scratching their extravagant beards, fretting whether a small raise might "bump" them from the freebie bracket to the taxpayer bracket. After all, an income of $799.99 incurred no tax, but an income of $800.01 did. Could a $0.02 raise really cost you $24 in tax?

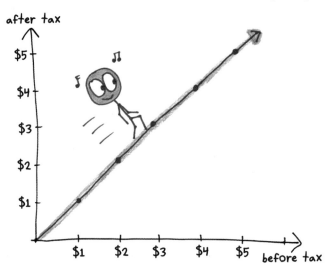

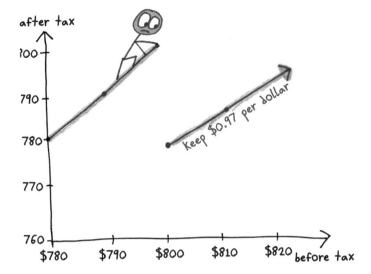

Thankfully, no. Like its modern equivalent, the tax didn't apply to all income—just on the margin, to the last dollar earned. Whether you were a poor Appalachian farmer or Cornelius Vanderbilt (the Bill Gates of the railroad age), your first $800 was tax-free.

If you made $801, then you'd pay the 3% on the last dollar alone, for a total liability of just $0.03.

If you made $900, then you'd pay the 3% on the last $100, for a total liability of just $3.

And so on.

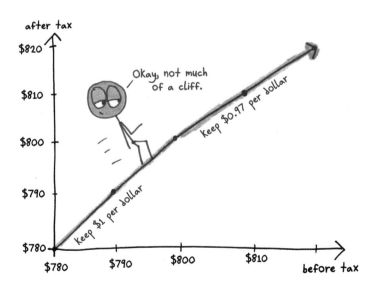

To visualize this, I like to imagine the government divides your money into buckets. The initial bucket fits $800 and is labeled "Survival Money." You fill this first, and pay no tax on it.

The second bucket is labeled "Leisure Money." Once the first bucket hits capacity, the rest of your money goes here. Of this, the government takes 3%.

By 1865, the government had raised the rates and brought in a third bucket:

Then the war ended. The income tax lay dormant for a few decades, until a flurry of activity at the end of the century. In 1893, a financial panic wracked the country. In 1894, the income tax came riding in to save the day. Then in 1895, the Supreme Court ruled that—ha ha, oops—that ban on "direct" taxes applies to you too, income tax. You're unconstitutional. How embarrassing.

It took two decades for a constitutional amendment to revive the tax. Even then, it returned without flags and fanfare, preferring to sneak back in on tiptoes. In 1913, the first year of the permanent income tax, only 2% of households paid it. The marginal rates maxed out at a measly 7%, which didn't even kick in until $500,000 (the inflation-adjusted equivalent of $11 million today).

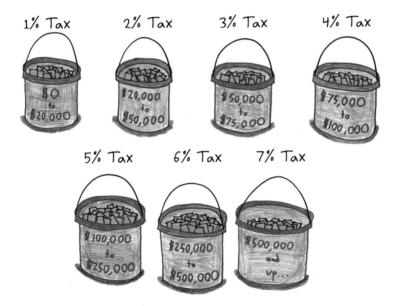

The rates have a pleasing ABC elegance, climbing from 1% to 7% in a tidy linear progression, with cutoffs at nice round numbers.

But why these in particular? No reason, honestly.

"Individual judgments will naturally differ," President Woodrow Wilson wrote, "with regard to the burden it is fair to lay upon incomes which run above the usual levels." No airtight mathematical formula can dictate the "right" rates. Rate setting is a subjective process, full of guesswork, politicking, and value judgments.

Of course, arbitrary legal lines are not unique to taxation. In the US, the day before your 21st birthday, you can't buy alcohol; the day after, you can. In theory, the law could ease you in gradually—allow you to buy beer at 19, wine at 20, and hard liquor at 21—but society has chosen a bright, simple line over a hard-to-enforce gradient.

The architects of that 1913 income tax went the other way. I find it funny that they broke such a modest tax into such tiny increments. Their seven brackets match the number we have today, except ours cover a range of 27%, and theirs just 6%. It seems that the government didn't trust the system's marginal nature, and thus sought to avoid big "jumps" between rates. Perhaps they were thinking in terms of this graph:

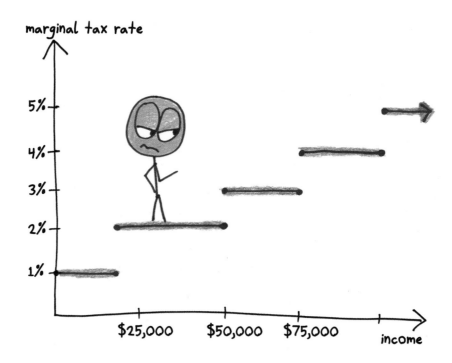

When you focus on the rates themselves, every bracket transition looks sharp and jagged. So lawmakers endeavored to smooth the system, to make it appear safe and gradual.

Wasted energy, if you ask me. Savvy taxpayers don't care about the abstract properties of the tax scheme. They care what they're paying—both overall and on the margin—and by design, the income tax ensures that earning an extra dollar will never alter the taxes already paid. The relevant graph is this one:

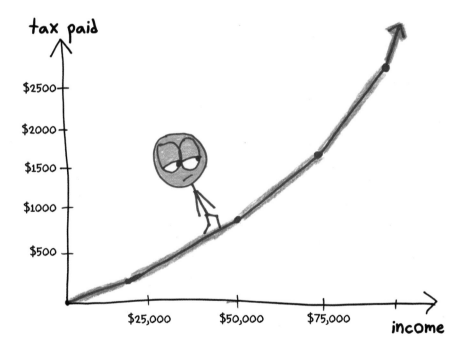

Whereas the first graph shows stark jumps, the second shows a single continuous line. It reflects the shifting rates via its changing slope: nice and shallow for a low tax rate, harsh and steep for a high one. This better captures how people actually experience bracket transitions: not as abrupt leaps (*discontinuities*, to a mathematician) but as moments where one slope gives way to another (*nondifferentiable points*, in calculus-speak).

Evidently, nobody told Congress, because they took their cliff wariness to a comical extreme in the 1918 tax scheme:

Income	Marginal Tax
0 to $4000	6%
$4000 to $5000	12%
$5000 to $6000	13%
$6000 to $8000	14%
$8000 to $10,000	15%
$10,000 to $12,000	16%

This is getting ridiculous!

(and so on... and so on... brackets $2000 wide, each 1% higher than before... until...)

$98,000 to $100,000	60%
$100,000 to $150,000	64%
$150,000 to $200,000	68%
$200,000 to $300,000	72%
$300,000 to $500,000	75%
$500,000 to $1 million	76%
$1 million and up	77%

This has gotten ridiculous.

Looking at this mess, you'll first notice that the rates have hulked out. In just five years, the marginal rate on America's wealthiest has grown elevenfold. I guess that's what happens when your country becomes embroiled in something called "the war to end all wars" (although the change would long outlast the conflict; the top rate hasn't dipped below 24% since).

Even more striking—to me, at least—is the sheer number of tax brackets instituted at that time. The US had more tax brackets than states, 56 to 48.

It reminds me of a favorite project from when I taught precalculus in California. Every year, I'd ask 11th graders to design their own income tax system. One industrious fellow named J. J. decided to take the "gradual transition" idea to its natural conclusion, designing a system where the marginal rates varied continuously, with no jumps whatsoever.

Imagine you want your marginal rates to start at zero and eventually reach 50% (on income above $1 million). You could do this with two brackets:

Or you could take the World War I–era approach, and break this into 50 brackets:

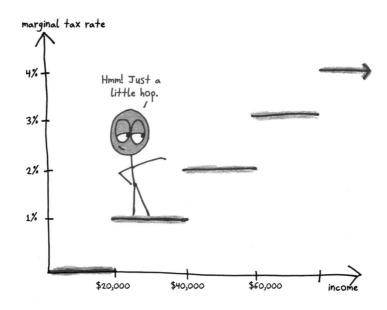

But why stop there? Make it 1000 brackets.

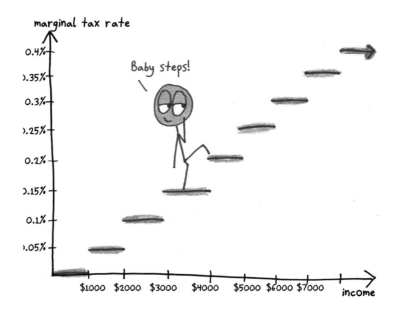

Or better yet, a million brackets.

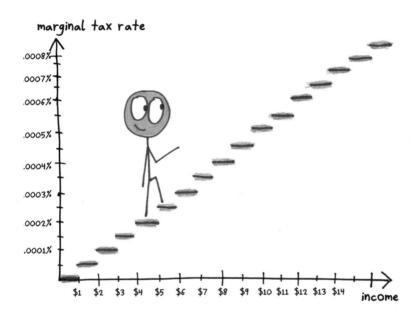

Heck, at the extreme, you could just connect your endpoints with a straight line.

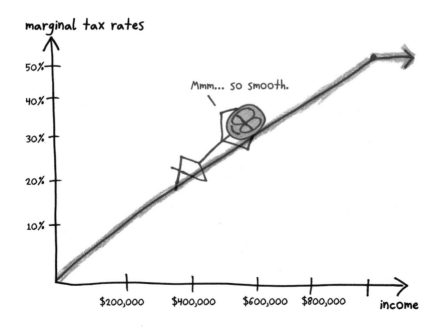

This final graph embodies a tax system where every fraction of a penny is taxed at an infinitesimally higher rate than the penny before. There is no such thing as "brackets"; each atom of income faces its own unique rate. Everywhere is a transition, and so, in some paradoxical sense, nowhere is.

Under such a system, the graph of total tax paid will look something like this:

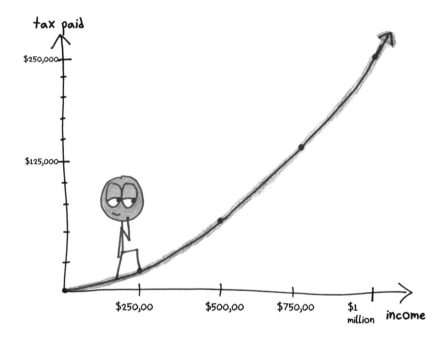

The slope grows gradually. No nondifferentiable points here. It's as smooth as a German car accelerating on the Autobahn.

Look at the income tax through US history, and you'll witness a similar acceleration. Over the first half of the 20th century, the tax grew from "unconstitutional proposal" to "tentative experiment" to "wartime necessity" to "the government's primary funding mechanism." In 1939, at the dawn of World War II, fewer than 4 million Americans paid income tax. By 1945, more than 40 million did. Revenue showed a similar growth, from $2.2 billion when the war began to $25.1 billion by its end. Today, its various state and federal forms amount to $4 trillion per year—almost a quarter of the country's economy.

As the income tax surged in 1942, the government commissioned Walt Disney to create a short film to inspire Americans to pay their taxes. The secretary of treasury demanded a fresh-faced custom-animated protagonist, but Walt Disney insisted on using the company's biggest star at the time. Thus, more than 60 million Americans got to see on their movie screens a display of patriotic, tax-paying fervor ("Oh boy! Taxes to beat the Axis!") from none other than Donald Duck.

The cartoon must have worked, because two years later, the top marginal tax rate achieved its historical high-water mark: 94%. Other countries pushed even higher. In the 1960s, the UK taxed top earners at a marginal rate of 96%, prompting the Beatles to begin their greatest album with the song "Taxman" and these scathing lyrics:

> *Let me tell you how it will be*
> *There's one for you, nineteen for me*

The most extraordinary tale of high marginal tax rates comes from—guess where?—Sweden. Rather than tell the tale myself, I shall defer to children's book author Astrid Lindgren, who in 1976 published a satire of her experiences, in the evening tabloid *Expressen*. It tells of a woman named Pomperipossa, living in a land called Monismania. Some folks in this land complained about the "oppressive taxes" that funded the welfare state. But Pomperipossa, despite her 83% marginal tax rate, did not. Instead, she contented herself with keeping 17%, and, "filled with joy," she "kept skipping down the road of life."

Now, Pomperipossa (a stand-in for author Lindgren) wrote children's books. In the eyes of the government, this made her a "small business owner," subjecting her to "social employer fees." But Pomperipossa did not understand the implications of this until a friend pointed them out:

> *"Are you aware of the fact that your marginal tax*
> *rate this year is 102%?"*

> *"You are talking nonsense," Pomperipossa said.*
> *"That many percent does not exist."*

> *For she was not particularly familiar with*
> *Higher Mathematics.*

This created an extraordinary scenario. For every dollar that Pomperipossa earned, she owed the government the entire dollar . . . plus an additional two cents. It was like the old nightmare of the "bracket bump" had come to life: higher earnings, lower wealth. Except instead of a singular cliff, after which normality returns, Pomperipossa had stepped through the looking glass, and now faced a never-ending downward slide. The more books she sold, the more impoverished she became.

> *"These terrible little children that are sitting in all the nooks and crannies of the world . . . how much money will their disastrous eagerness to read bring me this year?" . . . Big fat checks could mercilessly attack her when she least suspected it.*

Of the first $150,000 she earned, she would get to keep $42,000. But above that lay only heartbreak. Additional earnings would do worse than just vanish; they would take some of her earlier income along with them. Each additional $100,000 pretax would reduce her income by $2000 posttax.

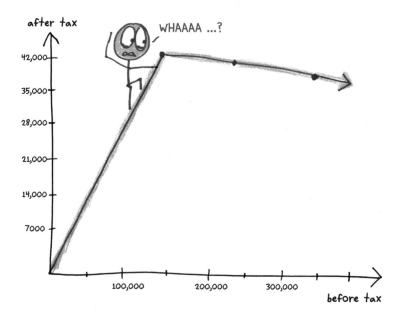

At worst, Pomperipossa figured, she'd earn $2 million. If that happened, she'd be left with a measly $5000. Pomperipossa couldn't believe it.

> *She said to herself: "My dear old woman . . . There are decimal points and all those things, surely you have counted wrong, there must be*

50,000 left for you." She started over, but the result did not change one bit . . . She now understood that it was something dirty and shameful to write books, since it was punished so severely.

The story ends with Pomperipossa, her earnings all sapped by taxes, going on welfare. The final line: "And she never, ever wrote any books again."

The figures in the story come directly from Lindgren's life, and make Sweden's top rate today—a mere 67%—look as quaint as a butter churn.

"Pomperipossa in Monismania" caught fire in Sweden, igniting a fierce debate that helped deal the Social Democratic Party its first electoral defeat in 40 years. Lindgren, a longtime Social Democratic supporter, kept voting for the party in spite of her grievances.

Back in the US, although the arguments surrounding the income tax are now a century old, they're as vibrant (and acrimonious) as ever.

My students' projects captured pretty much every side of the debate. Some tweaked the rates upward in the name of redistribution, or downward in the name of promoting economic growth. One student, who is as clever as he is cantankerous, designed a regressive system, whereby higher income levels faced a *lower* marginal tax rate. He argued that the poor needed an "incentive" to earn money. Maybe he was sincere, maybe he was satirizing Republican politics, or maybe he was just trolling me. (He got an A.) Still others sought a radical Robin Hood system of economic justice and chose top rates close to 100%. A few students chose rates *above* 100%, wishing to inflict Pomperipossa's fate on the ultrarich. Still other students wanted to step outside the income tax framework altogether, to envision a whole new regime.

In short, I saw inventiveness everywhere and consensus nowhere. I guess that's what they call "America."

Chapter 23

ONE STATE, TWO STATE, RED STATE, BLUE STATE

In America, we believe in government of the people, by the people, for the people. So it's a shame that the people never seem to agree on anything.

I find it hard enough to build consensus among a family choosing pizza toppings. And yet somehow, our noisy continent-spanning democracy must make collective yes-or-no decisions on even graver matters. We go to war, or we don't. We elect this president, or that one. We incarcerate Lakers fans, or let them walk the streets with impunity. How can we blend 300 million voices into a single national chorus?

Well, it demands some technical work. Taking a census. Totaling votes. Tabulating. Apportioning. This quantitative labor, if a little dry, is nevertheless vital. That's because, at its heart, representative democracy is an act of mathematics.

The simplest democratic system is called "majority rules." This creates a single tipping point, a threshold at 50%. There, one vote can flip you from "loser" to winner" (or vice versa).

But forget "simple." This is America! For electing our president, we've crafted a doozy of a mechanism known as the Electoral College, a mathematical and political oddity that introduces dozens of tipping points into the mix. On top of fueling our endless talk of "red states" and "blue states," it endows our elections with fascinating mathematical properties. In this chapter, I aim to follow these twists and turns in the story of the Electoral College:

1. It begins as a layer of human beings.
2. It becomes a layer of mathematics.
3. It transforms into a state-by-state "winner-take-all" system.
4. After all that, it behaves pretty much like a popular vote—
   but with a few fun surprises.

In the end, understanding the Electoral College comes down to a matter of marginal analysis. In the 300-million-legged race called US democracy, what is the meaning of a single step?

1. DEMOCRACY AS A GAME
OF TELEPHONE

In the summer of 1787, 55 dudes in wigs met daily in Philadelphia to hammer out a new plan for national government. Today, we call their design "the Constitution," and it sits alongside the cheesesteak at the foundation of our national character.

The Constitution outlined an elaborate system for determining the "President" (as in "one who presides"). He would be appointed every four years by special electors: "men chosen by the people for the special purpose, and at the particular conjuncture." I think of the electors as a one-time CEO-search committee. They convene for a single purpose, select the nation's new leader, and then disband.

Where would the electors come from? Well, the states. Each got one elector per congressman, and could select them by whatever method it pleased.

The logic was that local voters knew only local politicians. In a benighted time, decades before hashtags, how could citizens judge the oratory and haircuts of distant, unknown men? That's why the Electoral College filtered people's preferences through three steps: First, you vote for your local state representative; second, they choose electors (or create a system for their selection); and third, the electors determine the president.

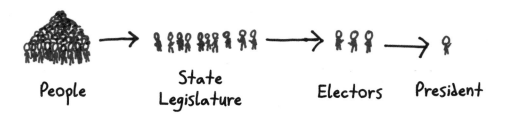

People   State Legislature   Electors   President

But by 1832, every state except South Carolina had tossed its electors into the democratic fray, choosing to let the people decide.

What changed?

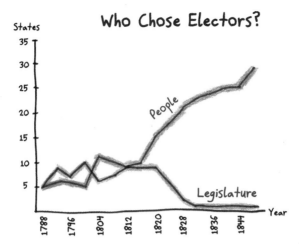

The Philly 55 had envisioned an enlightened democracy of conscience-driven statesmen, unencumbered by those nasty, rivalrous things called "political parties." Then they went out and formed things that looked an awful lot like political parties.

Hypocrisy aside, this had advantages. Now, you didn't need to study the conscience of every single candidate. You could just familiarize yourself with the party platforms and vote for the team you prefer. No need for an intermediary.

And so the country moved from human "electors" to abstract "electoral votes," the layer of statesmen replaced by a layer of mathematics.

Of course, as with many elements of the Constitution, this system encoded an advantage for slavery. How did it work? Well, recall that a state got one elector for each congressman. That included senators (two per state) and House members (varying in proportion to population; in 1804, they ranged from one to 22 per state).

Electors = Senators + House Members

The Senate half of the equation gave a boost to small states, by treating little Rhode Island (fewer than 100,000 people back in the day) on par with big Massachusetts (more than 400,000). But the real plot twist unfolded over in the House. The Philly 55 had debated whether, when apportioning House members, the tally of people represented ought to include slaves. If yes, then the slave-owning South would get extra representatives; if not, the North stood to benefit.

They compromised on counting them partway. Every five slaves would register as three people. Hence the most infamous fraction in legal history: 3/5.

The Electoral College imported this proslavery compromise into presidential elections, like a virus downloaded from an email attachment. Here's the elector count from 1800, compared to a hypothetical system where slaves went uncounted:

State	Electors	If Slaves Weren't Counted	
Virginia	24	− 5	
North Carolina	14	− 1	
Maryland	11	− 2	
South Carolina	10	− 1	South: −10
Kentucky	8		
Georgia	6	− 1	
Tennessee	5		
Pennsylvania	20	+ 2	
Massachusetts	19	+ 2	
New York	19	+ 1	
Connecticut	9	+ 1	North: +10
New Jersey	8	+ 1	
New Hampshire	7	+ 1	
Vermont	6	+ 1	
Rhode Island	4		
Delaware	3	+ 1	

Perhaps a 10-elector swing doesn't sound like much. It was. Of the country's first 36 years, it spent 32 led by slave-owning Virginians. The only interruption came from John Adams of Massachusetts, who lost a close reelection bid in 1800 . . . So close, in fact, that 10 electors would have swung it.

2. WHY "WINNER-TAKE-ALL" WON AND TOOK ALL

In his inaugural address after that bitter election of 1800, Thomas Jefferson struck a conciliatory note. He said that the two parties shared common principles, common dreams. "We are all Federalists," he said. "We are all Republicans."

But look at the Electoral College today, and you will not see a kumbaya country of yin-yang balance. You'll see a map filled with red and blue. California is all Democrats. Texas is all Republicans. There is no "we," only "us" and "them," like the highest-stakes two-player board game imaginable.

How did the Electoral College get here?

This is where a mathematician's nose starts twitching. Sure, we're letting the people vote, but how will those votes be aggregated and tabulated? What mathematical process will convert raw preferences into a final choice of electors?

Say you're modern-day Minnesota. You've got to boil 3 million votes down to 10 electors. How will you do it?

One option: assign electors **in proportion** to the actual votes. A candidate with 60% of the vote gets six electors. A candidate with 20% gets two. And so on.

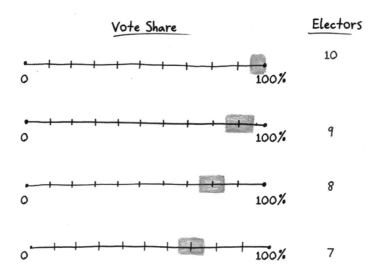

Though logical, this system has gained no traction. In the early days, states tried some weird stuff—in Tennessee, for example, voters chose county delegates who chose electors who chose the president—but none saw fit to go proportional.

An alternative: assign electors **by geography**. You've got 10 electors; why not break your state into 10 districts, and give one elector to the winner of each?

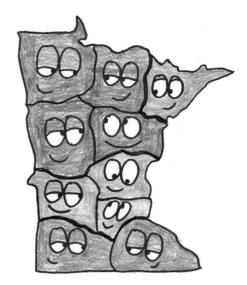

This system enjoyed a heyday in the 1790s and early 1800s. It has since gone extinct.

Its closest living relative is a version where you pick electors **by House district**. Then, because each state has two more electors than it has House seats, you give your final pair of electors to the winner of the statewide vote.

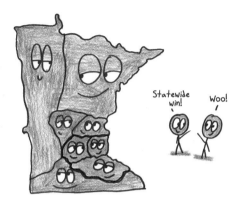

Today, this funky system is practiced in just two states: Nebraska and Maine. Nowhere else.

What the heck are the other 48 states doing? They follow a radical scheme: **winner-take-all**. Under this approach, the statewide winner gets every elector.

"Winner-take-all" carries a powerful implication: margin of victory doesn't matter. In 2000, George W. Bush won Florida by less than 600

votes. In his reelection, he won it by 400,000. But the larger margin wasn't any "better" for him than the razor-thin one. Winner-take-all collapses the continuous spread of possible percentages into two discrete outcomes, in a way the Philly 55 neither specified nor envisioned.

Why, then, is it practiced in 96% of states?

This is a problem straight out of game theory, the mathematics of strategy selection. To understand the outcome, we'll need to climb inside the mind of a state-level politician.

Let us begin in California. Democrats reign in the state legislature. And—surprise, surprise—Democrats tend to get more presidential votes, too. Imagine you've got two choices: go proportional, or go winner-take-all.

Well, if you're a Democrat, winner-take-all boosts your party. To do anything else is to gift a handful of electors to the Republicans. Why even consider it?

In Texas, the same logic holds, with the color scheme reversed. Winner-take-all locks up every elector for Republicans. Why relinquish precious electors to your opponent?

In theory, if all states went proportional, neither party would benefit. It would be like both sides in a duel laying down their guns. That's the upper-left box.

But it's not a stable equilibrium. Once your foe has dropped her weapon, you might as well grab yours back. If every state does the same, then we soon land in the lower-right box—which is, of course, where 96% of the country sits today.

Listening to a description of the Electoral College, you'd think that state borders matter. That to share a driver's license is to share a special kinship. But that's not how state legislators act. By choosing the all-or-nothing system, they do what's best for the party, even if it means marginalizing the state.

For example, here's how the last 10 elections have turned out in Texas:

Year	Winner
2016	Republican
2012	Republican
2008	Republican
2004	Republican
2000	Republican
1996	Republican
1992	Republican
1988	Republican
1984	Republican
1980	Republican

Am I such a one-note character?

Going winner-take-all amounts to posting a big sign at the border that reads, "Don't Worry What My Voters Think; We Always Go Republican!" Under winner-take-all, 55% is as good as 85%, and 45% is no better than 15%. Going winner-take-all means the election is over before it begins, and so neither party has any reason to tailor its policies to win over your citizens. Why waste resources where they will make no difference?

On the other hand, here's how those elections look if you allocate electors proportionally:

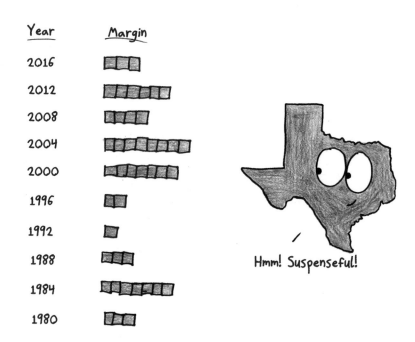

Year	Margin
2016	
2012	
2008	
2004	
2000	
1996	
1992	
1988	
1984	
1980	

Hmm! Suspenseful!

Now, modest shifts in the vote leave a real impact on the elector count. If Texas wants its voters to have a stake in the game, then it ought to go proportional, so that an extra Texan vote actually boosts a candidate's chances of winning. It would give campaigns reason to campaign.

Why doesn't this happen? Because "maximizing the impact of a marginal vote in my state" is not what motivates legislators. First and foremost, they are not Texans, or Californians, or Kansans, or Floridians, or Vermonsters.

They are all Democrats. They are all Republicans.

3. THE PARTISAN TIDES

After more than a century in which the Electoral College always aligned with the nationwide popular vote, we've seen two divergences in the last five elections. In 2000, the Democrats won the popular vote by 0.5%, while the Republicans won the Electoral College by five votes. Paper-thin margins on both fronts. In 2016, the disparity widened: the Democrats won the nationwide vote by 2.1%, while the Republicans won the Electoral College by 74.

Is the Electoral College now tilted to favor Republicans?

Nate Silver, the modern-day statistical prophet of the Electoral College, has a beautiful method for answering this question. It allows us to determine the Electoral College advantage, not just in exceptional years like 2000 and 2016, but in *any* election.

The procedure (which I'll apply to 2012) goes something like this:

1. **Line up the states from "reddest" to "bluest."** In 2012, that begins with Utah (a 48-point Republican victory), then Wyoming, then Oklahoma, then Idaho . . . all the way until Vermont, Hawaii, and finally DC (which went Democratic by 84 points).

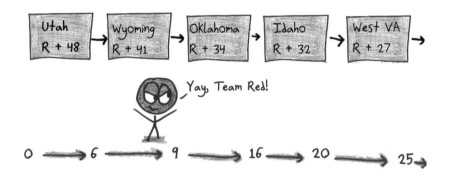

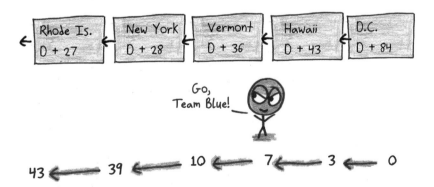

2. **Go to the middle and identify the "tipping point" state.** That's the state that pushes the winner past the 270-elector threshold required to win. In 2012, it was Colorado, which went Democrat by 5.4 points.

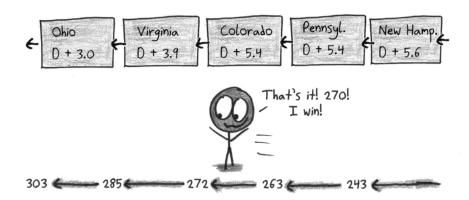

3. **Gradually lower the winner's vote share until the election is a virtual tie.** In reality, the Democrat won a comfortable nationwide victory. But in theory, he could have done 5.4% worse nationwide and still won the election, carrying Colorado by a single vote. So let's subtract 5.4% from his total in every state, to simulate a superclose election.

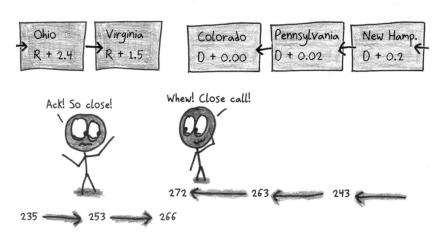

4. **This new, adjusted popular vote tells us the Electoral College advantage.**

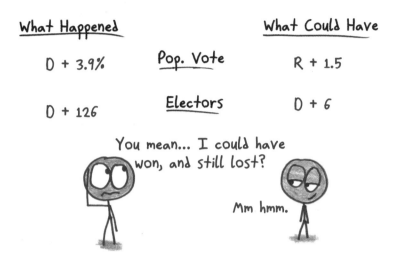

What Happened		What Could Have
D + 3.9%	<u>Pop. Vote</u>	R + 1.5
D + 126	<u>Electors</u>	D + 6

You mean... I could have won, and still lost?

Mm hmm.

Sometimes, the winning party has enjoyed an advantage that it didn't need (e.g., in 2008). Other times, it has managed to overcome an advantage of its opponent's (e.g., in 2004). What's most striking, though, is how the water level has risen and fallen across the years.

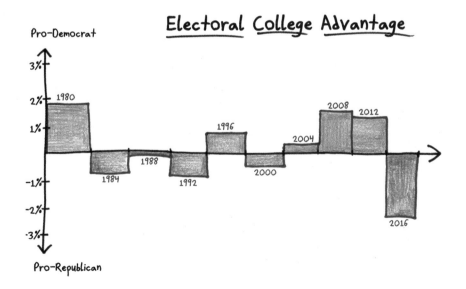

The last 10 elections have favored five Republicans and five Democrats. They average out to a Democratic advantage of less than 0.1%. Silver notes that "there's almost no correlation between which party has the Electoral College advantage in one election and which has it four years later. It can bounce back and forth based on relatively subtle changes in the electorate."

In this light, it seems that 2000 and 2016 really were flukes. In a nearby parallel universe, Republicans are fuming that Barack Obama won two elections in which he lost the popular vote, while Democrats gloat about their electoral invincibility.

What, then, to make of the Electoral College?

For all its mathematical complexity, it is more or less a randomizer. It echoes the popular vote, except when (for idiosyncratic reasons) it doesn't. There's no way to predict which direction the partisan advantage will tilt until November is almost upon us. That being the case, should we ditch the system?

I'm first and foremost a mathematician. That means I love elegance and simplicity—which the Electoral College does not have—but also wacky statistical scenarios, which it does.

I'm amused by a proposal called the National Popular Vote Interstate Compact, or, if you'd rather save nine syllables, the Amar Plan. The idea, crafted by two law professor brothers, is simple. States commit their electors to the winner of the nationwide popular vote—regardless of what happens within the state's borders. Thus far, the proposal has become law in 10 states and DC, totaling 165 electors. If enough states join to cross the 270-elector threshold, they'll gain control of the Electoral College. (In the meantime, the status quo reigns, since the laws are written not to take effect until a critical mass signs on.)

The Constitution lets states allocate electors as they wish. If they want a popular vote, it's their prerogative. In the peculiar history of the Electoral College, it would be another marginal step, a strange new cusp of history.

Chapter 24

THE CHAOS OF HISTORY

Y ou eye the title of this chapter with justified skepticism. "History?" you ask. "What do you know of history, mathematician?"

I mumble a few incoherencies about walruses, tax law, and some wigs in Philadelphia; your pity grows.

"Historians find causal patterns in the sprawl of the past," you explain. "Your tidy formulas and quaint quantitative models have no place in this messy human realm."

I hunch my shoulders and begin to sketch a graph, but you shush me. "Run home, mathematician!" you say. "Go, before you embarrass yourself!"

Alas, I lost my chance to avoid embarrassment the day I blogged my first stick figure, and so, in halting tones, I begin to tell my tale.

Ew. keep your math away from my history.

I can't help it! Math just spills everywhere!

1. A NOR'EASTER FROM A
ROUNDING ERROR

In the winter of 1961, two surprises hit the East Coast, more or less simultaneously.

First, in Washington, DC, 8 inches of snow fell on the eve of President Kennedy's inauguration. Anxious Southern drivers, possibly interpreting the snow as a signal of Armageddon, abandoned their cars by the thousands. Apocalyptic traffic jams ensued. The US Army Corps of Engineers managed to clear a path for the inaugural parade only by recruiting hundreds of dump trucks and flamethrowers to the cause.

It was, in short, chaos.

Second, back in Kennedy's home state of Massachusetts, a researcher named Edward Lorenz found something funny. Since the year before, he'd been developing a computer model of the weather. To begin, you entered some initial conditions. Then, the computer ran them through a set of equations. In the end, it delivered a printout of the resulting weather. It could use these results as the initial conditions for the next day, iterating the process to create months' worth of weather from a single starting point.

One day, Lorenz wanted to re-create an earlier weather sequence. One of his technicians retyped the inputs, rounding them a little (e.g., 0.506127 to 0.506). The tiny errors—smaller than what weather instruments could even detect—should have faded into the background. And yet, within weeks of simulated weather, the new sequence had diverged completely. A tiny adjustment had created a whole new chain of events.

It was, in short, chaos.

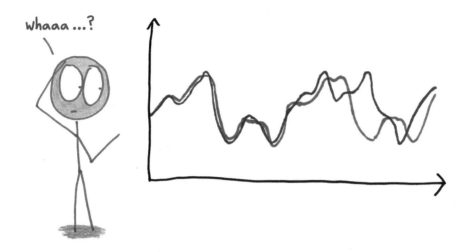

This moment marked the birth of a new experimental style of mathematics, an interdisciplinary insurgency that soon became known as "chaos theory." This field explored a variety of dynamic systems (gathering storms, turbulent fluids, populations in flux) with a strange set of shared features. They tended to follow simple and rigid laws. They were deterministic, with no room for chance or probability. And yet, because of the subtle interdependence of their parts, they defied prediction. These systems could amplify small changes into tremendous cascades, a tiny upstream ripple into an enormous downstream wave.

Lorenz and the nation's capital were both stunned by the unpredictability of the weather. But the link between the events runs deeper than this. Forget the chaos of the snowstorm, and consider instead the fact that John F. Kennedy was being inaugurated at all.

Three months earlier, he had defeated Richard Nixon in one of the closest elections in US history. He won the nationwide vote by just 0.17% and carried the Electoral College thanks to narrow margins in Illinois (9,000 votes) and Texas (46,000 votes). Half a century later, historians still debate whether voter fraud might have tipped Kennedy over the edge. (Verdict: Probably not, but who knows?) It is not hard to imagine a nearby parallel universe where Nixon eked out a win.

But it is very hard to imagine what would have happened next.

The Bay of Pigs invasion, the Cuban missile crisis, Kennedy's assassination, LBJ's ascension to the presidency, the Civil Rights Act, the Great Society, the Vietnam War, the Watergate break-in, Billy Joel's timeless hit "We Didn't Start the Fire" . . . All this and more hinged upon decisions made in the White House. A 0.2% swing in November of 1960 could have altered the path of world history, like a rounding error that births a nor'easter.

Ever since I was old enough to notice the world changing, I've been wondering how to conceptualize those changes. Civilization charts a path that's unknowable, unforeseeable, and unimaginable except after the fact. How can we make sense of a system where enormous and untold consequences can issue from a single indiscernible step?

2. TWO KINDS OF CLOCKWORK

"Ah, silly mathematician," you say. "You are being as skittish and alarmist as a DC driver in a slow flurry."

I gaze at you with wide Disney-sidekick eyes. Hey, I crave a predictable world as much as the next human.

"Human history is not chaos," you say. "It exhibits patterns and trends. Nations come and go. Political systems emerge and fade. Tyrants rise, accrue Instagram followers, and one day come tumbling down. All of this has happened before, and all of it will happen again."

I scratch my head for a moment and then reply with the tale of the pendulum.

$$\text{Length of Cycle} \approx 2 \times \sqrt{\text{Length of Pendulum}}$$

Mmm... sweet predictability.

Back at the dawn of the 17th century, when science first turned its bespectacled eye to the pendulum, it found a mechanism more reliable than any existing timepiece. The pendulum obeyed a simple equation: measure its length in meters, take the square root, double this number, and you'll arrive at the length of each cycle in seconds. There's a connection, then, between physical length and temporal length. A unity between space and time. Pretty cool.

The pendulum is what a mathematician calls "periodic," which means "repeating after a regular interval." It's like an undulating wave, or the rise and fall of the tides.

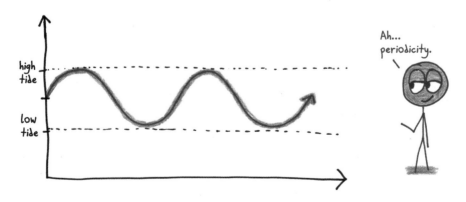

Ah... periodicity.

Sure, the pendulum suffers imperfections—friction, air resistance, fraying ropes—but these meaningless nudges and background noises don't spoil its reliability any more than a breeze unsettles a mountain. By the early 20th century, the finest pendulum-based clocks kept time accurate to within a second per year. That's why, to this day, the pendulum is a lovely mental token of a well-ordered universe.

But now for a plot twist: the double pendulum.

It's just a pendulum attached to another, still governed by the laws of physics, still described by a set of equations—so it ought to behave like its cousin, right? And yet, watch. Its swings are wild and erratic. It kicks left, gestures right, spins like a windmill, pauses to rest, and does it all again, except different.

What's going on here? It's not "random" in the mathematical sense. There are no cosmic dice, no quantum roulette wheels. It's a rule-bound, gravity-governed system. So why is its behavior so wacky? Why can't we predict its movements?

Well, in a word: sensitivity.

Release the double pendulum from one position and record its trajectory. Then release it from a nearly identical position, just a millimeter away, and watch as it follows a completely different path. In technical terms, the double pendulum is "sensitive to initial conditions." As with the weather, a tiny perturbation at the start can yield a dramatic transformation by the end. To make sound predictions, you'd need to measure its initial state with virtually infinite precision.

Now, does history strike you more as a single pendulum or a double one?

History kicks right; a tyrant topples. It swings left; a war begins. It pauses as if resting, and a startup scheme hatched in a California garage remakes the world in its emoji-drenched image. Human civilization is an interconnected system, hypersensitive to change, with brief intervals of stability and flurries of wild activity, deterministic yet wholly unpredictable.

In mathematics, an "aperiodic" system may repeat itself, but it does so without consistency. We're told that those who do not learn their history are doomed to repeat it, but perhaps the situation is worse than that. Maybe, no matter how many books we read and Ken Burns documentaries we watch, we are *all* doomed to repeat history—and still, in spite of ourselves, to be caught off guard, recognizing the repetitions only in hindsight.

3. THE GAME OF LIFE

"Come now, mathematician," you cajole. "People just aren't that complicated."

I frown.

"Don't take it personally," you say. "But you're not that hard to predict. Economists can model your financial choices. Psychologists can describe your cognitive shortcuts. Sociologists can characterize your sense of identity and pick apart your choice of Tinder photo. Sure, their colleagues in physics and chemistry may roll their eyes and make vomiting gestures, but the social scientists can achieve a surprising degree of accuracy. Human behavior is knowable, and history is the sum of human actions. So shouldn't it be knowable, too?"

That's when I grab a computer and show you the Game of Life.

Like all fun things, it involves a grid. Each square—known as a "cell"—can assume two states: "alive" or "dead." In fairness, the word "game" may overstate things, because Life unfolds from step to step according to four automatic, immutable rules:

1. If a dead cell has three living neighbors, it will come to life.

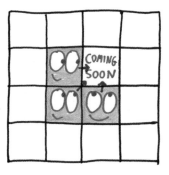

2. Otherwise, a dead cell will stage dead.

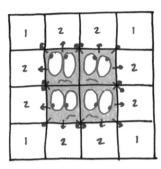

3. If a living cell has two or three living neighbors, it will stay alive.

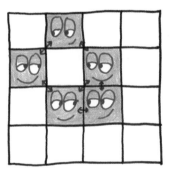

4. Otherwise, a living cell will die (either from loneliness or overcrowding).

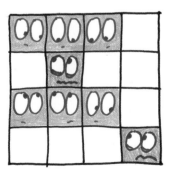

That's it. To start the game, just bring a few cells to life. Then, watch as the board changes, step by step, according to these rules. It's like a snowball that needs only a nudge to begin rolling downhill. No further input is required.

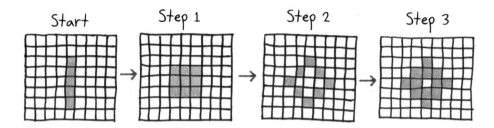

In this world, even more so than ours, earning a psychology degree is supereasy. Just memorize the four rules above, and you can now predict any cell's moment-to-moment behavior with flawless accuracy. Good work, doctor.

And yet, the board's long-term future remains opaque. Clusters of cells interact in subtle and hard-to-foresee ways. A pattern like the one below can give rise to never-ending growth; delete just one living cell from the original grid, and the growth will peter out.

Simplicity on a small scale gives way to strange emergent behavior on a large one.

I'm reminded of Amos Tversky's reply to the question of why he became a psychologist:

The big choices we make [in life] are practically random. The small choices probably tell us more about who we are. Which field we go into may depend on which high school teacher we happen to meet. Who we marry may depend on who happens to be around at the right time of life. On the other hand, the small decisions are very systematic. That I became a psychologist is probably not very revealing. What kind of psychologist I am may reflect deep traits.

In Tversky's view, small choices obey predictable causes. But large-scale events are the products of a devilishly complicated and interconnected system, in which every motion depends on context.

The same, I contend, is true of human history. A person is predictable; *people* are not. Their delicate interrelationships amplify some patterns and extinguish others, without any clear rhyme or reason.

It's no accident that the Game of Life, like Lorenz's weather simulation, comes from the computer age. By nature, chaos is not the kind of thing you can hold in your thoughts. Our minds are too inclined toward smoothing things over, too prone to round off the truth at a convenient decimal place. We needed brains bigger and faster than ours to make chaos tractable, to reveal its patterns—or its lack thereof.

4. NOT A BRANCH, BUT A THICKET

"Okay, mathematician," you sigh. "I see what you're getting at. What you're talking about is *alternate history*. Envisioning how civilization might have unfolded, if the past differed in some tiny yet consequential way."

I shrug.

"Well, you should have said so!" you exclaim, relieved. "Of course history's path sometimes forks—there are decisive battles, key elections, pivotal moments. That doesn't make history *chaotic*, just *contingent*. It's still subject to logic and causality. You needn't treat historical analysis as some kind of lost cause."

Turning Points of History, 1965 - 1970

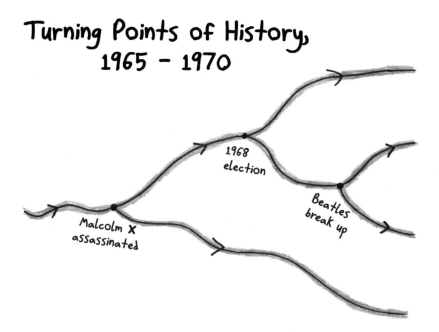

Now it's my turn to sigh and explain that, no, the problem runs deeper.

Start with a bomb—or rather, a pair of them. In 1945, the United States dropped two atom bombs on Japanese cities: Hiroshima (on August 6) and Nagasaki (on August 9). Alternate-history stories have wrestled with this moment. Take Kim Stanley Robinson's novella *The Lucky Strike*, in which the *Enola Gay*, rather than dropping the bomb on Hiroshima, is destroyed in a freak accident the day before, changing the course of the nuclear era.

Still, Robinson can tell only a single story, out of the trillions possible. What if the historic Japanese capital of Kyoto had remained atop the target list, as it did until late July? What if the August 9 weather over the city of Kokura had been clear instead of rainy, and the bomb intended for that city were never rerouted to Nagasaki? What if US president Harry Truman had realized before the bombing that Hiroshima was a civilian target, not (as he'd somehow believed) a strictly military one? There are more possibilities than any story can reckon with. Alternate history gives us branches; chaos warns that these have been pruned from a sprawling thicket.

Actual History

Idea of History

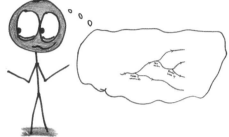

Even at their best, alternate histories can never reckon with the real nature of chaos. When pivotal importance is latent in every moment, linear storytelling becomes impossible. Take another of alternate history's favorite questions: What if the slaveholding Confederacy had won the American Civil War? It's a common topic for speculation; even Winston Churchill, not primarily known for his science fiction, weighed in. So, when television's HBO announced its plans for a show on this premise called *Confederate*, the writer Ta-Nehisi Coates—an expert both in African American history and in speculative fiction—issued an eloquent groan:

> *Confederate is a shockingly unoriginal idea, especially for the allegedly avant garde HBO. "What if the South had won?" may well be the most trodupon terrain in the field of American alternative history . . . [Consider] all the questions that are not being chosen: What if John Brown had succeeded? What if the Haitian Revolution had spread to the rest of the Americas? What if black soldiers had been enlisted at the onset of the Civil War? What if Native Americans had halted the advance of whites at the Mississippi?*

Alternate histories tend to dwell on the "great men" and "important battles" of conventional history, missing quieter possibilities that cut against the grain of the dominant culture. The rules of *plausibility* (what stories we find credible and compelling) don't always mirror those of *probability* (what things actually came close to happening).

True chaos is a narrative-destroying idea, a thought as anarchic as a bomb.

5. THE COASTLINE OF
WHAT WE KNOW

"Okay then, mathematician," you say, holding on to your patience by your fingernails. "You're saying that historical understanding is a sham. That the trends of history are aperiodic mirages. That our attempts to infer causality are doomed, because in a hyperconnected system like human civilization, where microscopic changes yield macroscopic effects, everything causes everything. That we can never, ever predict what will happen next."

"Well, when you put it that way, I sound like kind of a jerk," I say.

You glare. Point taken.

Well, if studying history is a fool's errand, than what exactly do you recommend?

Um... study history anyway?

Maybe human history is what chaos theorists call an "almost-intransitive" system. It looks pretty stable for long stretches and then, suddenly, shifts. Colonialism gives way to postcolonialism. Divine monarchy gives way to liberal democracy. Liberal democracy gives way to corporate-run anarcho-capitalism. That kind of thing. Even if historians can't show us what lies beyond the next cusp, they can at least characterize the shifts that came before and shed light on the current state of human affairs.

Chaos counsels humility. It teaches us, again and again, the limits of what we can know.

In 1967, legendary rascal and chaos mathematician Benoit Mandelbrot published a brief bombshell of a paper titled "How Long Is the Coast of Britain?" The problem is harder than it sounds, because—weird as it seems—the length of Britain's coastline depends on how you measure it.

Start with a ruler 10 kilometers long, and you'll measure a certain length.

Then zoom in and switch to a 1-kilometer ruler. Stretches that previously appeared straight now reveal themselves to be quite jagged. Your shorter ruler can fit into these nooks and crannies, and so you arrive at a longer total length.

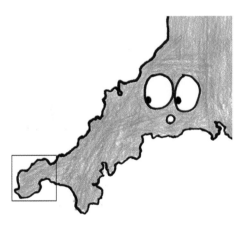

We're not done. Switch to a 100-meter ruler, and the same process will repeat. Curves and crinkles overlooked by the longer rulers now become apparent, increasing the total length.

We can keep repeating this process. The closer you look, the longer the coastline becomes—in theory, forever.

Suffice it to say, this is kind of weird. With most inquiries, looking closer helps clarify the answer. Here, in a frightening inversion, looking closer causes the problem to unravel. It never simplifies, never resolves.

This trend reaches its extreme with the Koch snowflake, a mathematical object consisting of bumps upon bumps upon bumps. Although it occupies only a small patch of the page, the length of its boundary is theoretically infinite.

Perimeter: **3 cm**   Perimeter: **4 cm**   Perimeter: **$5\frac{1}{3}$ cm**   Perimeter: **∞ ?**

In the graphic novel *From Hell*—a work of speculative history, about the series of murders in London's Whitechapel neighborhood in 1888—writer Alan Moore brings in Koch's snowflake as a metaphor for the nature of historical scholarship. "Each new book," he says in the afterword, "provides fresh details, finer crenellations of the subject's edge. Its area, however, can't extend past the initial circle: Autumn 1888, Whitechapel."

In Moore's telling, there is something bottomless about the study of history. The further we zoom, the more we see. A finite stretch of time and space can enclose infinite layers of detail, allow for endless chains of analysis. Chaos is complexity all the way down, never pixelating, never terminating, never resolving.

Is history chaotic like the Game of Life—simple on a small scale yet unpredictable on a large one? Or is its unpredictability like the weather's—with wild swings on the small day-to-day scale yet averaging out in the long run to a stable climate? Or is it perhaps like the Koch snowflake—with chaos on every level, complexity at every scale? These metaphors compete in my mind, like three PowerPoint presentations projected onto a single screen. I sometimes feel I'm on the cusp of figuring it out—and then I check the news, and the world has changed again, into yet another strange and unknowable shape.

ENDNOTES

ASIDES, SOURCES, THANK-YOUS,
MATHEMATICAL DETAILS, AND JOKES THAT WERE
TOO WEIRD AND DISRUPTIVE FOR THE TEXT

Section I: How to Think Like a Mathematician

Chapter 1: Ultimate Tic-Tac-Toe

- **Their board looked like this:** This game's origins are murky. It may have first appeared in *Games* magazine in the late 1990s or early 2000s (although the folks at *Games* hadn't heard of it when I asked them). In 2009, a wooden board-game version called Tic-Tac-Ku won an award from Mensa. Perhaps the game had multiple independent discoveries, like dancing or calculus.

- **with my students:** When I first showed the game to my students at Oakland Charter High School in 2012, they took to calling it Ultimate Tic-Tac-Toe. My 2013 blog post with this title seems to have marked an inflection point for the game's popularity: Wikipedia, a handful of academic papers, and a larger handful of phone apps all refer to the game by that name. The obvious conclusion: take pride, Matadors! You named this thing.

- **"what does any of this have to do with math?":** I thank Mike Thornton, who read an early draft of this chapter and asked exactly this question. Mike's edits are like Leonard Cohen's songwriting or Hemingway's prose: I always knew they were good, but the older I get, the more I appreciate them.

- **deeper insight into the nature of rectangles:** The key idea is that thin rectangles have disproportionately large perimeters, whereas squarish rectangles have disproportionately large areas. So just pick a long thin rectangle (e.g., 10 by 1) and a squarish one (e.g., 3 by 4).

- **Again, see the endnotes:** If you require the solutions to use only whole numbers, the problem is a lot of fun. Here's my derivation of a formula that generates an infinite family of solutions:

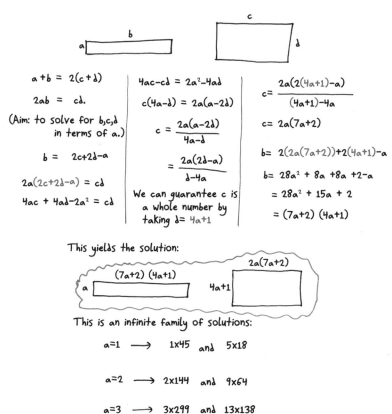

$$a + b = 2(c + d)$$

$$2ab = cd.$$

(Aim: to solve for b, c, d in terms of a.)

$$b = 2c + 2d - a$$

$$2a(2c + 2d - a) = cd$$

$$4ac + 4ad - 2a^2 = cd$$

$$4ac - cd = 2a^2 - 4ad$$

$$c(4a - d) = 2a(a - 2d)$$

$$c = \frac{2a(a - 2d)}{4a - d}$$

$$= \frac{2a(2d - a)}{d - 4a}$$

We can guarantee c is a whole number by taking $d = 4a + 1$

$$c = \frac{2a(2(4a+1) - a)}{(4a+1) - 4a}$$

$$c = 2a(7a + 2)$$

$$b = 2(2a(7a+2)) + 2(4a+1) - a$$

$$b = 28a^2 + 8a + 8a + 2 - a$$

$$= 28a^2 + 15a + 2$$

$$= (7a + 2)(4a + 1)$$

This yields the solution:

$$a \quad (7a+2)(4a+1)$$

$$4a+1 \quad 2a(7a+2)$$

This is an infinite family of solutions:

$$a = 1 \longrightarrow 1 \times 45 \text{ and } 5 \times 18$$

$$a = 2 \longrightarrow 2 \times 144 \text{ and } 9 \times 64$$

$$a = 3 \longrightarrow 3 \times 299 \text{ and } 13 \times 138$$

and so on!

That gives you *infinite* solutions, but not *all* of them, because other choices for *d* may still yield integer values for *c*. For example, this formula misses my favorite solution: 1 by 33 and 11 by 6. My colleague Tim Cross, a dexterous hand with Diophantine equations, showed me a nifty way to characterize all possible integer solutions. In the misanthropic tradition of my field, I'll leave that as an "exercise for the reader."

• **a guaranteed winning strategy:** The actual strategy is a little too complicated to describe here, but you can see it implemented by some folks from Khan Academy at https://www.khanacademy.org/computer-programming/in-tic-tac-toe-ception-perfect/1681243068.

• **a clever Brit locked himself in an attic:** I recommend the full story: Simon Singh, *Fermat's Last Theorem* (London: Fourth Estate Limited, 1997).

• **the great Italian mathematician Cardano:** This quote comes from the only dictionary I've ever read for pleasure: David Wells, *The Penguin Dictionary of Curious and Interesting Numbers* (London: Penguin Books, 1997).

Chapter 3: What Does Math Look Like to Mathematicians?

• **I zip through five chapters of a *Twilight* novel:** I'm more of a *Hunger Games* guy, to be honest.

• **mathematicians employ strategies:** Michael Pershan, the Most Curious and Analytical Man in the World, articulated this "strategy" idea before I had gotten there myself. My thanks for his help with this chapter.

• **peeling the circle like an onion:** There's a nice animation at www.geogebra.org/m/WFbyhq9d.

Chapter 4: How Science and Math See Each Other

• **forgetting their spouse's names on occasion:** I've been married to a mathematician for five years. So far, she seems to remember me.

• **the twisted history of knot theory:** For more, see: Matt Parker, *Things to Make and Do in the Fourth Dimension* (London: Penguin Random House, 2014).

• **a shifting landscape of strange curvatures:** Thanks to Matthew Francis and Andrew Stacey for their help on this point. I wanted to assert that the universe is "hyperbolic" or "elliptic" as opposed to "Euclidean," but they informed me that the real picture is a hard-to-fathom patchwork of these simpler geometries.

Stacey said: "Riemannian geometry generalises [*sic; he's not American, but don't hold it against him*] Euclidean geometry in many ways and is far richer, but a few things get lost on the way, primarily anything that relies on saying how things relate at different points in space." That includes the concept of "parallel."

Francis adds this interesting historical wrinkle: "William Kingdon Clifford proposed using non-Euclidean geometry to replace forces in the 19th century, but he never got much farther than 'this would be neat.' It wouldn't surprise me if others were thinking those thoughts too." Einstein, of course, collaborated closely with mathematicians; no breakthrough stands alone.

• **folks like Bertrand Russell:** Experience this story in graphic novel form: Apostolos Doxiadis et al., *Logicomix: An Epic Search for Truth* (New York: Bloomsbury, 2009).

• **One logician's mother:** James Gleick, *The Information: A History, a Theory, a Flood* (New York: Knopf Doubleday, 2011). Stellar book. This delightful fact comes from page 213.

• **"unreasonable effectiveness":** Eugene Wigner, "The Unreasonable Effectiveness of Mathematics in the Natural Sciences," Richard Courant lecture in mathematical sciences delivered at New York University, May 11, 1959, *Communications on Pure and Applied Mathematics* 13 (1960): 1–14. It's a cracking essay.

Chapter 5: Good Mathematician vs. Great Mathematician

• **teachers like me:** My own teacher for this chapter was David Klumpp, whose feedback combines the erudition of someone like Carl Sagan with the gentle humanity of someone like Carl Sagan. (David Klumpp is Carl Sagan.)

• **algebra was perhaps the dullest:** Israel Kleiner, "Emmy Noether and the Advent of Abstract Algebra," *A History of Abstract Algebra* (Boston: Birkhäuser, 2007), 91–102, https://link.springer.com/chapter/10.1007/978-0-8176-4685-1_6#page-2. I've done some real violence to Kleiner's argument; the key point is that the 19th century brought big advances in analysis and geometry, while algebra remained in a more concrete and primordial state.

a mathemetician named Emmy Noether: Joaquin Navarro, *Women in Maths: From Hypatia to Emmy Noether. Everything is Mathematical.* (Spain: R.B.A. Coleccionables, S.A., 2013).

• **"She often lapsed":** Professor Grace Shover Quinn, as quoted in: Marlow Anderson, Victor Katz, and Robin Wilson, *Who Gave You the Epsilon? And Other Tales of Mathematical History* (Washington, DC: Mathematical Association of America, 2009).

• **Sylvia Serfaty:** All the Serfaty quotes come from her interview with: Siobhan Roberts, "In Mathematics, 'You Cannot Be Lied To,'" *Quanta Magazine*, February 21, 2017, https://www.quantamagazine.org/sylvia-serfaty-on-mathematical-truth-and-frustration-20170221. I recommend Roberts's writing on math the way I recommend R.E.M.'s best albums: zealously.

• **Grothendieck:** Colin McLarty, "The Rising Sea: Grothendieck on Simplicity and Generality," May 24, 2003, http://www.landsburg.com/grothendieck/mclarty1.pdf.

• **Gaussian correlation inequality:** Natalie Wolchover, "A Long-Sought Proof, Found and Almost Lost," *Quanta Magazine*, March 28, 2017, https://www.quantamagazine.org/statistician-proves-gaussian-correlation-inequality-20170328. The story is a great read, even after you've encountered my spoilers of it.

• **a beloved physics teacher:** Farhad Riahi, 1939–2011.

• **A student named Vianney:** Whereas Corey is a pseudonym, Vianney is not. I figure she deserves the victory lap. I've fudged the dialogue to fill gaps of memory, but this is pretty much how it went down.

• **Ngô Bầu Châu:** The quotes come from a press conference Châu gave at the 2016 Heidelberg Laureate Forum. My enormous thanks to A+ human being Wylder Green and the HLF team for making my attendance there possible.

Section II: Design

• **this will never pan out:** The interior angle of a regular pentagon is 108°. If you try to arrange three around each vertex, you'll have 36° left over, which isn't enough room for a fourth. The only regular polygons that can tile a plane are triangles (60°), squares (90°), and hexagons (120°), because you need an angle that goes evenly into 360°.

• **Geometry's laws are not like that:** There is *some* flexibility. In my examples, I've taken Euclid's assumption about parallels; you could take another. But once you do, every other rule will follow by logical necessity.

Why relegate this important caveat to an obscure endnote? Well, I figure anyone finicky enough to question whether we're in Euclidean space is finicky enough to read endnotes, too.

Chapter 6: We Built This City on Triangles

• **the Great Pyramid of Giza:** I should credit here the essential source for this chapter: Mario Salvadori, *Why Buildings Stand Up* (New York: W. W. Norton, 1980). It's a stately, upstanding book, without which this chapter would have toppled like an erroneous proof. My thanks also to Will Wong, architect of thought and of intramural sports victories, for his help with this chapter.

• **a peculiar loop of rope:** I learned about the Egyptian rope pullers from: Kitty Ferguson,

Pythagoras: His Lives and the Legacy of a Rational Universe (London: Icon Books, 2010).

• **frustums:** Truncated pyramids with trapezoids for lateral faces. A word worth knowing.

- **less than 0.1% of its internal volume:** Wikipedia lists dimensions for three passageways (descending, ascending, horizontal) and three rooms (king's chamber, queen's chamber, grand gallery). These volumes totaled to 1340 m³, which in a structure of 2,600,000 m³ amounts to roughly 0.05%. I rounded this up to 0.1%, then (with more thanks to Wikipedia) calculated the volume of the Empire State Building (2,800,000 m³). Taking 0.1% of this, you get 2800 m³. Dividing by the area of a single floor (roughly 7400 m²), you get a height of 38 centimeters, which is about 15 inches. I rounded up to 2 feet. Then, just as I completed this manuscript, a hidden chamber was discovered! Still, my rounding should more than cover the possible error.

- **crucial supporting character:** Pun 100% intended.
- **the beam's effect:** This discussion is pilfered from Salvadori's *Why Buildings Stand Up*, and no doubt loses much in the retelling.
- **Architects are no fools:** Ted Mosby notwithstanding.
- **hence, the term "I-beam":** I borrow again from Salvadori. As Will Wong points out, a more traditional presentation would foreground the desirable properties (stress distribution, torque prevention, etc.) created by the cross-section's I shape.
- **The Pratt truss:** My knowledge of trusses derives from that greatest of human creations, Wikipedia. For more, check out the "Truss" and "Truss Bridge" pages.

Chapter 7: Irrational Paper

- **When I moved to England:** Thanks to Caroline Gillow and James Butler, two people with hearts so wide they make the Atlantic look piddling, for their help and encouragement on this chapter, and for making that "I moved to England" experience as wonderful as it was.
- **My 12-year-old student Aadam:** The same day Aadam coined "disintegers," an 11-year-old student named Harry responded to my greeting

of "Hello, algebraists!" with this reply: "Why not call us alge-zebras?" I tell ya, teaching is a good gig.
- **all exactly the same proportion:** Another delight is that each sheet of A1 has an area of precisely 1 m², and each step down the series cuts the area in half. Thus, eight sheets of A4 will give you precisely a square meter (except not *precisely*, because . . . irrationals).

Chapter 8: The Square-Cube Fables

- **Size matters:** While I was working on this chapter, my colleague-slash-role-model Richard Bridges pointed me to a delightful presentation of the same ideas written 80 years earlier. I went on to borrow liberally from that essay: J. B. S. Haldane, "On Being the Right Size," March 1926, https://irl.cs.ucla.edu/papers/right-size.pdf.
- **the only available pan has double the dimensions:** I'm ignoring height here, since when baking brownies, you don't fill the pan up to the brim.
- **local sculptor Chares:** I learned this story from Kitty Ferguson's *Pythagoras*. Like all good fables, it is most likely apocryphal.
- **It was three-dimensional:** John Cowan—whom I thank for fact-checking this chapter, and for wielding his omniscience in the friendliest, least scary way possible—adds this wrinkle: "In fact, the Colossus of Rhodes, like the Statue of Liberty, was hollow. It was made of bronze plates with iron reinforcing rods. Consequently, as the height grew by n, the cost grew only by n^2." Still too much for poor Chares.
- **There's no such thing as giants:** I first learned this in college in Laurie Santos's psychology course

Sex, Evolution, and Human Nature. I mean, I already knew there weren't giants, but Professor Santos's explanation of why (which I now see was perhaps inspired by Haldane) helped form the genesis of this chapter.
- **the overwhelming weight of his torso:** Please, call your senators and urge them to invest in our critical Dwayne Johnson infrastructure before it is too late.
- **why bricks plummet and paper flutters:** The mathematics of air resistance underlies another quick fable: "Why Big Sailboats Need *Huge* Sails." When you double your boat's dimensions, the sails' area (2D) quadruples, but the ship's weight (3D) octuples. You'll be catching proportionally less wind. Thus, a double-length ship needs roughly triple-height sails.
- **the way I fear them:** John Cowan adds: "Another point about ants is that they don't bleed, because being so small they are 'all surface' and don't need internal fluid to distribute oxygen to the interior." In surface-heavy creatures, diffusion suffices.
- **babies are small:** See www.thechalkface.net/resources/baby_surface_area.pdf. This vignette

draws inspiration from a math teacher who goes by the nickname "The Chalk Face."

- **square-cube fables are everywhere:** Here are a few extras that didn't fit in the chapter.

  #1: Why are big hot-air balloons more cost-effective? Because the canvas required depends on the surface area (2D), while the lift provided depends on the volume of helium (3D).

  #2: Why do huge dino-birds like turkeys take so long to cook? Because heat absorption grows with surface area (2D), but heat required grows with volume (3D).

  #3: Why is dry wheat perfectly safe while wheat dust is an explosive hazard? Because exothermic chemical reactions occur on the surface of a substance, and tiny dust particles are far more surface-heavy than intact wheat stalks.

- **"paradox of the dark night sky":** I first learned about Olber's paradox from Pieter van Dokkum's, astronomy course Galaxies and Cosmology. If you want someone to credit (or blame) for this chapter's eclecticism, look no further than my liberal arts education.

- **brighten you to death:** I give John Cowan the final word here: "What if there are Dark Things (planets, dust, etc.) between us and the stars? Wouldn't that make some stars invisible and eliminate the paradox? No, because the stars behind the Dark Things would over time heat them up to the same temperature as the stars, eliminating all darkness."

- **Edgar Allan Poe prose poem:** E. A. Poe, *Eureka: A Prose Poem* (New York: G. P. Putnam, 1848).

Chapter 9: The Game of Dice

- **these Roman emperors:** The historical facts in this chapter come from three sources, listed here in rough order of how much I plundered from them:

  - Deborah J. Bennett, *Randomness* (Cambridge, MA: Harvard University Press, 1998).
  - "Rollin' Bones: The History of Dice," *Neatorama*, August 18, 2014. Reprinted from the book *Uncle John's Unsinkable Bathroom Reader*, http://www.neatorama.com/2014/08/18/Rollin-Bones-The-History-of-Dice/.
  - Martin Gardner, "Dice," in *Mathematical Magic Show* (Washington, DC: Mathematical Association of America, 1989), 251–62.

- **does not make a fair die:** The snub disphenoid has two cousins, where every face is an equilateral triangle yet the die is unfair: (1) the triaugmented triangular prism (shown to me by Laurence Rackham), and (2) the gyroelongated square bipyramid (shown to me by Tim Cross and Peter Ollis). If you prefer quadrilateral faces, there's (3) the pseudo-deltoidal icositetrahedron (shown to me by Alexandre Muñiz).

triaugmented triangular prism

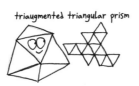

gyroelongated square bipyramid

pseudo-deltoidal icositetrahedron

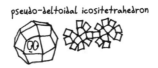

- **why aren't they more popular?:** To be fair, there are examples. Ancients in the Indus Valley threw triangular long dice molded from clay. Their contemporaries in India rolled rectangular ones, carved from ivory.

- **demands a red carpet:** This perhaps explains why all the modern long dice I've seen are either (a) tiny, or (b) twisted so as to keep the long faces equivalent while generating a more satisfying and contained tumble. The Dice Lab, whose work helped to inspire this chapter, sells some nice instances of the latter.

- **mathematical principle: *continuity*:** For more on this line of argument, see: Persi Diaconis and Joseph B. Keller, "Fair Dice," *American Mathematical Monthly* 96, no. 4 (April 1989): 337–39, http://statweb.stanford.edu/~cgates/PERSI/papers/fairdice.pdf.

- **this approach bugs people:** Nevertheless, several civilizations have done similar. Ancient Inuits rolled chair-shaped ivory dice, on which they counted only three of the six sides. The Papago Indians threw bison bones, counting only two of the four sides.

- **people try to cheat:** For ingenious and evil details, see: John Scarne, *Scarne on Dice*, 8th ed. (Chatsworth, CA: Wilshire Book Company, 1992).

- **Renumber the faces:** A related approach, unsubtle yet elegant, comes from the gangster Big Jule in the musical *Guys and Dolls*. His lucky dice are blank, with no spots on them at all. But don't worry: Big Jule remembers where all the spots *used* to be.

• **nothing appears to be wrong:** Well . . . almost. A true expert can recognize a tap, because from certain angles the faces do not follow the usual configuration, but rather a mirror image. For this reason and others, taps con artists always move their cheat

dice in and out of play quickly, to avoid detection.

• **to roll that d100:** As Ralph Morrison, knower of all things trivial and nontrivial, pointed out to me, the Dice Lab sells a gorgeous 120-sided die. For more: Siobhan Roberts, "The Dice You Never Knew You Needed," Elements, *New Yorker*, April 26, 2016, https://www.newyorker.com/tech /elements/the-dice-you-never-knew-you-needed.

Chapter 10: An Oral History of the Death Star

• **the team responsible:** In reality, I drew heavily from: Ryder Windham, Chris Reiff, and Chris Trevas, *Death Star Owner's Technical Manual* (London: Haynes, 2013). I wrote this chapter in part to bring a smile to the face of Neil Shepherd, but don't tell him I said that.

• **architects, engineers, and grand moffs:** If you enjoy this kind of thing, I recommend: Trent Moore, "Death Star Architect Defends Exhaust Ports in Hilarious Star Wars 'Open Letter,'" *SyFy Wire*, February 14, 2014, http://www.syfy.com/syfywire /death-star-architect-defends-exhaust-ports -hilarious-star-wars-open-letter%E2%80%99.

• **face of a dianoga:** My thanks go to Gregor Nazarian, who makes my heart soar like a John Williams theme, and who helped greatly to improve this chapter. (I've tagged him here because he suggested I use the dianoga, not because his face resembles one.)

• **at a diameter of about 400 kilometers:** The rock and ice figures come from Megan Whewell of the National Space Centre, as quoted by: Jonathan O'Callaghan, "What Is the Minimum Size a Celestial Body Can Become a Sphere?" *Space Answers*, October 20, 2012, https://www.spaceanswers.com /deep-space/what-is-the-minimum-size-a-celestial

-body-can-become-a-sphere/. The figure for "imperial steel" is my own blind extrapolation.

• **only 140 kilometers across:** My research turned up figures ranging from 120 kilometers to 150 kilometers.

• **2.1 million people on the Death Star:** I searched and searched to find a more realistic figure, but the sources all seem to agree. *Star Wars* fans have debated this question (https://scifi .stackexchange.com/questions/36238/what-is-the -total-number-of-people-killed-on-the-2-death-stars -when-they-explode) and tend to arrive at figures between 1 and 2 million. Windham et al. give roughly 1.2 million people and 400,000 droids. Wikipedia gives 1.7 million people and 400,000 droids. To prove the robustness of my point, I've taken the highest numbers I could find.

• **exploiting a thermal vent:** I'm not sure that I've successfully shoehorned this narrative into the new continuity established by *Rogue One*. No doubt certain stubborn and admirable purists will be upset by the liberties here taken. On the other hand, I've given these characters from "a galaxy far, far away" access to 2010 census data on West Virginia, so maybe redistributing the architectural credit isn't my worst crime against canon.

Section III: Probability

Chapter 11: The 10 People You Meet in Line for the Lottery

• **People earning at least $90,000 are likelier:** Zac Auter, "About Half of Americans Play State Lotteries," *Gallup News*, July 22, 2016, http://news .gallup.com/poll/193874/half-americans-play-state -lotteries.aspx. Even so, it's still true that the lottery is a "regressive tax," because when a poor and a rich person spend the same amount of money, the poor person is spending a larger percentage of their income.

• **my home state of Massachusetts:** Paige DiFiore, "15 States Where People Spend the Most on

Lotto Tickets," Credit.com, July 13, 2017, http://blog .credit.com/2017/07/15-states-where-people-spend -the-most-on-lotto-tickets-176857/. Though the rankings fluctuate from year to year, Massachusetts has been at or near the top since my birth in 1987.

• **$10,000 Bonus Cash:** I drew these odds from MassLottery.com: http://www.masslottery.com/games /instant/1-dollar/10k-bonus-cash-142-2017.html.

• **Put "10,000" and "bonus" in front of any word:** Try it. 10,000 bonus tortillas; 10,000 bonus fist bumps; 10,000 bonus puppy snuggles. It's foolproof.

• **20% of our tickets are winners:** Almost half of these merely cover the $1 cost of the ticket, so "winner" is perhaps less apt than "nonloser."

• **Poll Americans about why they buy lottery tickets:** Charles Clotfelter and Philip Cook, "On the Economics of State Lotteries," *Journal of Economic Perspectives* 4, no. 4 (Autumn 1990): 105–19, http://www.walkerd.people.cofc.edu/360/AcademicArticles/ClotfelterCookLottery.pdf.

• **when states introduce new lottery games, overall sales rise:** Kent Grote and Victor Matheson, "The Economics of Lotteries: A Survey of the Literature," College of the Holy Cross, Department of Economics Faculty Research Series, Paper No. 11-09, August 2011, http://college.holycross.edu/RePEc/hcx/Grote-Matheson_LiteratureReview.pdf. Also, I offer my enormous gratitude to Victor Matheson for taking the time to look this chapter over.

• **the UK National Lottery draw offered:** Alex Bellos, "There's Never Been a Better Day to Play the Lottery, Mathematically Speaking," *Guardian*, January 9, 2016, https://www.theguardian.com/science/2016/jan/09/national-lottery-lotto-drawing-odds-of-winning-maths.

• **11% of lottery drawings fit the bill:** Victor Matheson and Kent Grote, "In Search of a Fair Bet in the Lottery," College of the Holy Cross Economics Department Working Papers. Paper 105, 2004, http://crossworks.holycross.edu/econ_working_papers/105/.

• **from losing too much:** For example, a syndicate led by Stefan Klincewicz bought up 80% of the possible tickets for an Irish National Lottery in 1990. His team wound up splitting the top prize but turned a profit thanks to the generous secondary prizes. Klincewicz told reporters that he avoids the UK lottery because it lacks those smaller prizes. Source: Rebecca Fowler, "How to Make a Killing on the Lottery," *Independent*, January 4, 1996, http://www.independent.co.uk/news/how-to-make-a-killing-on-the-lottery-1322272.html.

• **the 1992 Virginia State Lottery:** I first learned about this story in one of the best recent pop math books: Jordan Ellenberg, *How Not to Be Wrong* (New York: Penguin Books, 2014). I then tracked its trajectory through a trio of delicious old news stories: (1) "Group Invests $5 Million to Hedge Bets in Lottery," *New York Times*, February 25, 1992, http://www.nytimes.com/1992/02/25/us/group-invests-5-million-to-hedge-bets-in-lottery.html; (2) "Group's Lottery Payout is Postponed in Virginia," *New York Times*, March 7, 1992, http://www.nytimes.com/1992/03/07/us/group-s-lottery-payout-is-postponed-in-virginia.html; and (3) John F. Harris, "Australians Luck Out in Va. Lottery," *Washington Post*, March 10, 1992, https://www.washingtonpost.com/archive/politics/1992/03/10/australians-luck-out-in-va-lottery/cbbfbd0c-0c7d-4faa-bf55-95bd6590dc70/?utm_term=.9d8bd00915e8.

• **the late 1600s:** Anne L. Murphy, "Lotteries in the 1690s: Investment or Gamble?" University of Leicester, dissertation research, http://uhra.herts.ac.uk/bitstream/handle/2299/6283/905632.pdf?sequence=1. I love the names of 17th-century English lotteries like "the Honest Proposal" and "the Honourable Undertaking." They might as well be called "Yes, We Know We Could Swindle You, but We Promise We Won't."

• **a fun would-you-rather:** If you're the kind of well-read person who skims the endnotes, then you surely know it already, but I shall nevertheless cite: Daniel Kahneman, *Thinking Fast and Slow* (New York: Farrar, Straus and Giroux, 2011).

• **prospect theory:** Daniel Kahneman and Amos Tversky, "Prospect Theory: An Analysis of Decision Under Risk," *Econometrica* 47, no. 2 (1979): 263, http://www.princeton.edu/~kahneman/docs/Publications/prospect_theory.pdf.

• **far likelier to be poor:** Clotfelter and Cook, "On the Economics of State Lotteries."

• **a profit of $30 billion per year:** Derek Thompson, "Lotteries: America's $70 Billion Shame," *Atlantic*, May 11, 2015, https://www.theatlantic.com/business/archive/2015/05/lotteries-americas-70-billion-shame/392870/. See also: Mona Chalabi, "What Percentage of State Lottery Money Goes to the State?" FiveThirtyEight, November 10, 2014, https://fivethirtyeight.com/features/what-percentage-of-state-lottery-money-goes-to-the-state/.

• **I suspect it's hypocrisy:** Back in the 1790s, French revolutionaries considered the lottery an evil excess of the monarchic state. But they hesitated to abolish it after seizing power, for the simple reason that they needed the money. How else can you turn a taxphobic citizenry into dutiful taxpayers? Source: Gerald Willmann, "The History of Lotteries," Department of Economics, Stanford University, August 3, 1999, http://willmann.com/~gerald/history.pdf.

• **rip-off of a popular illegal game:** Clotfelter and Cook.

• **post-USSR Russians:** Willmann, "The History of Lotteries."

• **so little in winnings:** Bingo averages a payout of $0.74 in winnings for every dollar spent. Horseracing pays $0.81. Slot machines pay $0.89. State lotteries average around $0.50. Source: Clotfelter and Cook.

• **This tradition goes way back:** Willmann.

• **offset, dollar for dollar:** Grote and Matheson, "The Economics of Lotteries."

- **John Oliver quips:** "The Lottery," *Last Week Tonight with John Oliver*, HBO, published to YouTube on November 9, 2014, https://www.youtube.com/watch?v=9PK-netuhHA.
- **a chance to fantasize about winning money:** Why do middle-aged people buy lottery tickets at higher rates? I'm just spitballing, but perhaps it's because middle age is the best time to be a Dreamer. Young adults can imagine other paths to fortune. Older adults are beyond such longings. Only the middle-aged are old enough to realize that no magic financial transformation is forthcoming, but young enough to crave it.
- **winning a lottery jackpot tends to make people less happy:** For a full and up-to-date discussion, see: Bourree Lam, "What Becomes of Lottery Winners?" *Atlantic*, January 12, 2016, https://www.theatlantic.com/business/archive/2016/01/lottery-winners-research/423543/.
- **next socioeconomic stratum:** See, for example: Milton Friedman and L. J. Savage, "The Utility Analysis of Choices Involving Risk," *Journal of Political Economy* 56, no. 4 (August 1948): 279–304, http://www2.econ.iastate.edu/classes/econ642/babcock/friedman%20and%20%20savage.pdf.
- **instant games, with their modest top prizes:** Grote and Matheson, "The Economics of Lotteries."
- **odd economy of scale:** My example is adapted from Clotfelter and Cook.

Chapter 12: Children of the Coin

- **I was pressed into teaching 10th-grade biology:** Witness this typical scene from April 2010:

    KISA (with bright-eyed curiosity): What exactly happens in the endoplasmic reticulum?

    ME: There's no way of knowing. It is an unsolvable enigma, beyond human imagination.

    TIM (in a bored monotone): The textbook says that's where proteins are folded.

    ME: Well, obviously, Tim. I meant aside from that.
- **The riddles of resemblance strike at the heart of biology:** For a take that's more sophisticated than mine but still plenty readable, see: Razib Khan, "Why Siblings Differ Differently," Gene Expression, *Discover*, February 3, 2011, http://blogs.discovermagazine.com/gnxp/2011/02/why-siblings-differ-differently/#.Wk7hKGinHOi.
- **unfolding in very few ways:** This same logic underlies the concept of entropy, the universe's tendency toward disorder.

    Take a collection of bricks. There are precious few ways to form a building, and many dull, indistinguishable ways to form a pile of rubble. Over time, random changes will accumulate, almost all of them making your arrangement more rubble-y, and almost none making it more building-y. Thus, time favors rubble over bricks.

    Similarly, there are very few ways for the particles of food dye to gather on one side of a water glass; it'd be like the molecules all flipping heads. But there are many, many ways for those particles to scatter more or less evenly through the liquid; each such scattering is like a different combination of heads and tails. That's why random processes lead in erratic but inexorable ways toward greater entropy, an even mixing of the universe's ingredients. The cosmic preference for disorder is, at its heart, combinatorial.

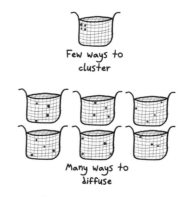

Few ways to cluster

Many ways to diffuse

- **between 16 and 30 heads:** This has a probability around 96%, so for one in 25 readers, my prophecy will prove false. That said, if as many as 25 readers actually go flip 46 coins, then the math-book-reading audience is even more hard-core than I suspected.
- **Blaine Bettinger:** The graph here comes straight from him. See more at: https://thegeneticgenealogist.com/wp-content/uploads/2016/06/Shared-cM-Project-Version-2-UPDATED-1.pdf.

    The *x*-axis is labeled "centimorgans," which are (to me, anyway) science's most confusing unit. A centimorgan is a length of chromosome that has a 1% chance of being broken up by a chromosomal crossover in any given generation. With close relatives, you'll share lots; with distant relatives, very few. Thus, "centimorgans shared" is a measure of genetic proximity.

    So far, so good. But because crossing over happens with different probabilities all over the genome, a centimorgan is not a constant length. Where crossing over is common, centimorgans are short; where it's rare, they're long. On top of that, different DNA-sequencing companies divide the human genome into different numbers of centimorgans. And did I mention that 100 centimorgans do *not* make a morgan?

A further puzzle: when I converted the centimorgans in this graph to percentages, I found the distribution centered not around 50%, but 75%. Why? My wife, Taryn, explained: It's because commercial DNA kits can't distinguish between having *both* chromosomes in common and having just *one*. Each counts as "matching." By the logic of coin flips, two siblings will tend to have 50% of their DNA single-matching,

25% double-matching, and 25% not matching at all. Thus, 75% will match singly or doubly. Hence, the graph centered on 75%.

- **roughly twice per chromosome:** The number I found was 1.6. It's higher for women than for men. In any case, my tripling is a gross underestimate of possible genomes, because crossing over can happen (in theory) at any point along the DNA sequence, which allows for myriad additional possibilities. For more, see Ron Milo and Rob Phillips, "What Is the Rate of Recombination?" *Cell Biology by the Numbers*, http://book.bionumbers .org/what-is-the-rate-of-recombination/.

Chapter 13: What Does Probability Mean in Your Profession?

- **Actual Probability vs. How People Treat It:** Adapted from Kahneman, *Thinking Fast and Slow*, 315.
- **Donald Trump defeated Hillary Clinton:** For more on this outcome, see 99.997% of all content published on social media since.
- **your job was pretty boring:** Michael Lewis,

Liar's Poker: Rising Through the Wreckage on Wall Street (New York: W. W. Norton, 1989).
- **You're increasing the odds:** Nate Silver, *The Signal and the Noise: Why So Many Predictions Fail—but Some Don't* (New York: Penguin Books, 2012), 135–37.

Chapter 14: Weird Insurance

- **insurance is a strange and silly thing:** For this chapter, I drew tremendous help from Shizhou Chen, my former 10th-grade student and current intellectual superior. The original draft had a more presumptuous title and a less self-aware intro; Shizhou cut me down to size. "Quirky and funny examples," she wrote, "but not exactly the full picture of what insurance is about." Touché, Shizhou.
- **China five millennia ago:** Emmett J. Vaughan, *Risk Management* (Hoboken, NJ: John Wiley & Sons, 1996), 5.
- **"Iran, 400 BCE":** Mohammad Sadegh Nazmi Afshar, "Insurance in Ancient Iran," *Gardeshgary, Quarterly Magazine* 4, no. 12 (Spring 2002): 14–16, https://web.archive.org/web/20080404093756/; http:// www.iran-law.com/article.php3?id_article=61.
- **National Lottery Contingency Scheme:** "Lottery Insurance," This Is Money, July 17, 1999, http:// www.thisismoney.co.uk/money/mortgageshome /article-1580017/Lottery-insurance.html.
- **What might surprise you is the margin:** Shizhou makes a good point here: In niche insurance markets like this, there's less competition, and thus higher markups than in big markets like dental or home insurance.
- **if you can't insure 'em, join 'em:** Shizhou makes this astute addendum: "Small business owner? Yup. Large firm? Totally illegal. The account-

ing department would rather put $1 million under 'insurance' than $5 under 'lottery.'"
- **"multiple-birth insurance":** Laura Harding and Julian Knight, "A Comfort Blanket to Cling to in Case You're Carrying Twins," *Independent*, April 12, 2008, http://www.independent.co.uk/money/insurance /a-comfort-blanket-to-cling-to-in-case-youre-carrying -twins-808328.html. The quote from David Kuo also comes from this piece.
- **financial and psychological:** For similar analysis, see: "Insurance: A Tax on People Who Are Bad at Math?" *Mr. Money Mustache*, June 2, 2011, https:// www.mrmoneymustache.com/2011/06/02/insurance -a-tax-on-people-who-are-bad-at-math/. As the Mustache man writes: "Insurance of all types—car, house, jewelry, health, life—is a crazy field swayed by lots of marketing, fear, and doubt."
- **alien-abduction policy:** Check it out at http://www.ufo2001.com. My source: Vicki Haddock, "Don't Sweat Alien Threat," *San Francisco Examiner*, October 18, 1998, http://www.sfgate .com/news/article/Don-t-sweat-alien-threat -3063424.php.
- **selling abduction coverage in earnest:** Teresa Hunter, "Do You Really Need Alien Insurance?" *Telegraph*, June 28, 2000, http://www .telegraph.co.uk/finance/4456101/Do-you-really -need-alien-insurance.html.

• **the following criteria:** These are my own invention. Shizhou lent me her notes from an undergraduate module, and the list of "characteristics of an insurable risk" differed slightly:

1. "Potential loss is significant enough that people would exchange premium for coverage."
2. "Loss and its economic value is well-defined and out of the policyholder's control."
3. "Covered losses reasonably independent among policyholders."

• **Hole-in-One Prize Insurance:** Two such companies are Insurevents (http://www.insurevents.com/prize.htm) and National Hole-in-One (http://holeinoneinsurance.co.uk).

• **Jordan's Furniture ran a glitzy promotion:** Scott Mayerowitz, "After Sox Win, Sofas Are Free," ABC News, October 29, 2007, http://abcnews.go.com/Business/PersonalFinance/story?id=3771803&page=1.

• **the company Wedsure:** See its policies at http://www.wedsure.com.

• **reimbursing for weddings canceled:** Haddock, "Don't Sweat Alien Threat."

• **began offering to reimburse third-party:** Amy Sohn, "You've Canceled the Wedding, Now the Aftermath," *New York Times*, May 19, 2016, https://www.nytimes.com/2016/05/22/fashion/weddings/canceled-weddings-what-to-do.html.

• **Become so expert in the risk:** This helps broaden your business, too, by allowing you to serve as a sort of risk consultant. "Special expertise is one of the reasons firms buy insurance," Shizhou told me. "When filming movies, there's always an insurance inspector to make sure actors are safe. Without them, movies would have more mad and pointless explosions than what we see today."

• **"reinsurance":** Olufemi Ayankoya, "The Relevance of Mathematics in Insurance Industry," paper presented February 2015.

• **insure against career-ending injury:** "Loss-of-Value White Paper: Insurance Programs to Protect Future Earnings," NCAA.org, http://www.ncaa.org/about/resources/insurance/loss-value-white-paper.

• **purchasing an additional plan:** Andy Staples, "Man Coverage: How Loss-of-Value Policies Work and Why They're Becoming More Common," SportsIllustrated.com, January 18, 2016, https://www.si.com/college-football/2016/01/18/why-loss-value-insurance-policies-becoming-more-common.

• **Jake Butt:** A great name, right? Will Brinson, "2017 NFL Draft: Jake Butt Goes to Broncos, Reportedly Gets $500K Insurance Payday," CBS Sports, April 29, 2017, https://www.cbssports.com/nfl/news/2017-nfl-draft-jake-butt-goes-to-broncos-reportedly-gets-500k-insurance-payday/.

• **health insurance:** You could fill a bookshelf with this stuff. I recommend the work of *Vox* journalist Sarah Kliff, https://www.vox.com/authors/sarah-kliff. Shizhou recommends the two-part *This American Life* episode: "More Is Less" (#391), and "Someone Else's Money" (#392), http://hw3.thisamericanlife.org/archive/favorites/topical.

Chapter 15: How to Break the Economy with a Single Pair of Dice

• **What is the basic activity of Wall Street banks?:** Far and away the most essential source for this chapter was: David Orrell and Paul Wilmott, *The Money Formula: Dodgy Finance, Pseudo Science, and How Mathematicians Took Over the Markets* (Hoboken, NJ: John Wiley & Sons, 2017).

• **make educated guesses:** Okay, in my case, it's more like "make up random probabilities." Wall Street employs two more serious approaches. First, they anchor on historical data. And second, they look at the market price for similar bonds and use this to infer what the probability of default must be. This latter method can create spooky dependencies and feedback loops: Instead of exercising your own judgment, you echo the perceived wisdom of the market. Put a pin in that idea.

• **collateralized debt obligations:** Michael Lewis, *The Big Short: Inside the Doomsday Machine* (New York: W. W. Norton, 2010). Or, if you prefer listening to reading: "The Giant Pool of Money," *This Ameri-can Life*, episode #355, May 9, 2008A. This episode launched the essential *Planet Money* podcast.

• **surrealist painter René Magritte**: I saw these sketches in Belgium's lovely Musée Magritte on June 4, 2017. I recommend a visit if you're in Brussels and would like some surrealism to go with your frites.

• **an *idiosyncratic* risk or a *systematic* one:** These are the standard finance terms, brought to my attention by Jessica Jeffers, to whom I owe a huge thank-you for her help with this chapter. I am only 37% joking when I say that Jess is my choice for Fed chair.

• **the notorious "Gaussian copula":** Another essential source for this chapter, and this discussion in particular, is: Felix Salmon, "Recipe for Disaster: The Formula That Killed Wall Street." *Wired*, February 23, 2009, https://www.wired.com/2009/02/wp-quant/.

• **CDS. It stood for "credit default swap":** Also, "complete damn stupidity."

- **How could Wall Street be so stupid?:** See another essential source for this chapter: Keith Hennessey, Douglas Holtz-Eakin, and Bill Thomas, "Dissenting Statement," Financial Crisis Inquiry Commission, January 2011, https://fcic-static.law .stanford.edu/cdn_media/fcic-reports/fcic_final _report_hennessey_holtz-eakin_thomas_dissent.pdf.

- **"the wisdom of the crowd":** James Surowiecki, *The Wisdom of Crowds: Why the Many Are Smarter than the Few and How Collective Wisdom Shapes Business, Economies, Societies, and Nations* (New York: Anchor Books, 2004).
- **1987 stock crash:** Orrell and Wilmott, 54.

Section IV: Statistics

- **A survey of medical professionals:** Kahneman, *Thinking Fast and Slow*, 228.

Chapter 16: Why Not to Trust Statistics

- **citizens that H. G. Wells envisioned:** Thanks to Richard Bridges for (1) his help with this chapter; and (2) being a Platonist, a pragmatist, a teacher, and a brilliant mind, and for proving by construction that all those things can coexist.
- **Take household income:** All data from Wikipedia. For you, dear reader, only the best.
- **the tasty £2-per-jar tomato sauce:** Loyd Grossman's. They make a solid jar of tikka masala, too.
- **take a square root at the end:** My students tend to find it peculiar (and needlessly complicated) to square the distances, average them, and then take the square root. Why not just average the distances and leave it at that? Well, you're welcome to do this; the result is called the "mean absolute deviation," and it serves much the same role as standard deviation. However, it lacks some nice theoretical properties. Variances can be added and multiplied with ease; this makes them vital for building statistical models.
- **how the correlation coefficient *actually* works:** Okay, buckle up! Things are about to get hectical, which is my portmanteau of "hectic" and "technical." To begin, take a scatter plot—say, height against weight. Represent each person as a dot.

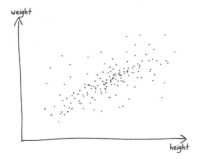

Now, find the *average height* and *average weight* in the population.

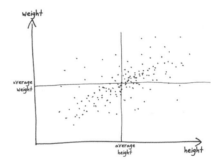

Next, take an individual person. How far is that person from the average height? How far from the average weight? Count "above average" as positive and "below average" as negative.

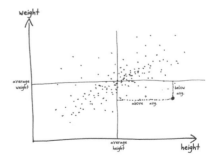

Then—and this is the crucial step—multiply these two values.

If the person is above average in both, you'll get a positive result. Same if they're below average in both (because the product of two negatives is positive). But if they're above average in one and below average in the other, then you'll get a negative result (because the product of a negative and a positive is negative).

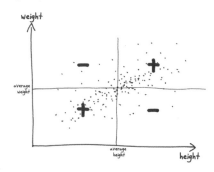

You do this for each and every person, then average together all the products. The result is something

called "covariance" (a cousin of the regular variance).

You're almost done! As a last step, divide this number, to yield a final result between -1 and 1.

(Divide by what? Well, consider the shortcoming of covariance: If people's weights and heights are both very spread out, then "distance from the average" will typically be a large number. In other words, covariance is larger for volatile variables and lower for stable ones, regardless of the relationship between them. How do we fix this problem? Just divide by the variances themselves.)

Whew! Now for the easy part: interpreting the values.

A positive value (e.g., 0.8) suggests that people above average on one variable (e.g., height) are usually above average on the other (i.e., weight). A negative value (e.g., -0.8) suggests the opposite: it's a population dominated by tall, light people and short, heavy ones. Finally, a value close to zero suggests there's no meaningful relationship at all.

Chapter 17: The Last .400 Hitter

- **an Englishman named Henry Chadwick:** As quoted in: "Henry Chadwick," *National Baseball Hall of Fame*, http://baseballhall.org/hof/chadwick-henry. "Every action," he said of baseball, "is as swift as a seabird's flight." As my friend Ben Miller points out, this raises a question: African swallow, or European?
- **all-time record is 400:** This record belongs to West Indies batsman Brian Lara, who in 2004 scored exactly 400 runs against England without making an out. I'm grateful for the strange statistical rhyme with this chapter's title.
- **"one true criterion":** As quoted in: Michael Lewis, *Moneyball: The Art of Winning an Unfair Game* (New York: W. W. Norton, 2003), 70. It will surprise no one when I confess that this chapter owes an unpayable debt to *Moneyball*, and if you find the story of baseball statistics tolerable (never mind interesting), then you'll enjoy the book.
- **you'd have to reset the ball at center field every five seconds:** I took the English Premier League's 38-game season. Calling each game 90 minutes, plus a generous 10 minutes for stoppage time, that's 3800 minutes. Twelve data points per minute (i.e., one per 5 seconds) would give 45,600 data points—still less than a baseball season's 48,000, but close enough.
- **the first edition of Ernest Hemingway's:** Ernest Hemingway, *The Old Man and the Sea*, *Life*, September 1, 1952. Above the title, it read: "The editors of Life proudly present for the first time and in full a great new book by a great American writer."

- **"Goodby to Some Old Baseball Ideas":** Branch Rickey, "Goodby to Some Old Baseball Ideas," *Life*, August 2, 1954. The subhead reads, "'The Brain' of the Game Unveils Formula That Statistically Disproves Cherished Myths and Demonstrates What Really Wins."
- **in 1858, "called strikes" were born:** E. Miklich, "Evolution of 19th Century Baseball Rules," 19cBaseball.com, http://www.19cbaseball.com/rules.html.
- **Enough such "balls":** It took a long time to settle on how *many* balls constituted a walk. At first it took three, then nine, then eight, then six, then seven, then five, until in 1889 the number finally settled at its current value of four.
- **Walks weren't deemed an official statistic until 1910:** Even then, writers didn't exactly embrace them. Just listen to sportswriter Francis Richter:

  The figures [for walks] are of no special value or importance . . . bases on balls are solely charged to the pitcher, beyond the control of the batsman, and therefore not properly to be considered in connection with his individual work, except as they may indicate in a vague way his ability to 'wait out' or 'work' the pitcher.

- Source: Bill James, *The New Bill James Historical Baseball Abstract* (New York: Free Press, 2001), 104.
- **hitters walk 18% or 19% of the time:** In 2017, for example, the league leader was Joey Votto, who

walked 134 times in 707 plate appearances. That's 19% of the time.

- **just 2% or 3%:** In 2017, Alcides Escobar walked 15 times in 629 plate appearances, for a rate of 2.4%. Tim Anderson outdid him by walking 13 times in 606 plate appearances, or 2.1%.

- **OPS:** As a math teacher, I've always hated this junkyard stat, because I lose life force every time someone adds two fractions with different denominators. I always wished they'd calculate a new statistic, "bases per plate appearance"—just like SLG, except you count walks on par with singles. Researching this chapter, I realized my folly: Although cleaner in concept, this new statistic would actually be less predictive in practice. For the 2017 numbers, its correlation with a team's run scoring is 0.873. That's worse than OBP.

- **On the 50th anniversary of the *Life* article:** Alan Schwarz, "Looking Beyond Batting Average," *New York Times*, August 1, 2004, http://www .nytimes.com/2004/08/01/sports/keeping-score -looking-beyond-batting-average.html.

- **The baseball dialogue in *The Old Man and the Sea*:** For example, "'I would like to take the great DiMaggio fishing,' the old man said. 'They say his father was a fisherman.'"

- **writer Bill James:** Scott Gray, *The Mind of Bill James: How a Complete Outsider Changed Baseball* (New York: Three Rivers Press, 2006). The book has some choice Bill James aphorisms, including: "There will always be people who are ahead of the curve, and people who are behind the curve. But knowledge moves the curve." Also: "When you add hard, solid facts to a discussion, it changes that discussion in far-reaching ways."

- **batting average is a sideshow:** And yet, through the 1970s, official MLB statistics listed team offenses in order of batting average, not runs scored.

"It should be obvious," quipped James, "that the purpose of an offense is not to compile a high batting average."

- **salaries soared:** Michael Haupert, "MLB's Annual Salary Leaders Since 1874," *Outside the Lines* (Fall 2012), Society for American Baseball Research, http://sabr.org/research/mlbs-annual -salary-leaders-1874-2012. I cheat big-time in the graph by not adjusting for inflation; for example, Joe DiMaggio's $90,000 salary in 1951 would be more like $800,000 in 2017 dollars. Even so, it's hard to argue with the impact of free agency.

- **"Baseball is the only field of endeavor":** Pete Palmer, *The 2006 ESPN Baseball Encyclopedia*, (New York: Sterling, 2006), 5.

- **a heartbreaking .39955:** Bill Nowlin, "The Day Ted Williams Became the Last .400 Hitter in Baseball," *The National Pastime* (2013), Society for American Baseball Research, https://sabr.org/research /day-ted-williams-became-last-400-hitter-baseball.

- **no one has hit .400 since:** Ben Miller (a gentleman, a culinary hero, and an incurable Sox fan) points out that Williams's success went far beyond that number. In 1941, he "had an OBP of .553, which stood as the single-season record for more than 60 years . . . Also, Ted Williams's career OBP of .482 is the highest ever. Oh, also, Ted Williams's career batting average of .344 is 6th-highest all-time, and you have to go all the way to Tony Gwynn at #17 to find the next all-time BA leader who played after 1940." No matter how you slice it, the Splendid Splinter was good at his job.

- **"If I had known hitting .400":** Bill Pennington, "Ted Williams's .406 Is More Than a Number," *New York Times*, September 17, 2011, http://www.nytimes.com/2011/09/18/sports/baseball /ted-williamss-406-average-is-more-than-a-number .html.

Chapter 18: Barbarians at the Gate of Science

- **we are living through an epoch of crisis:** A widely cited 2011 paper demonstrated the dangers of the standard statistical techniques by reaching an absurd conclusion: that listening to the Beatles song "When I'm Sixty Four" made students younger. Not that they *felt* younger. They were, chronologically, younger—or so the statistics said. The paper is cheeky, brilliant, and well worth reading: Joseph Simmons, Leif D. Nelson, and Uri Simonsohn, "False-Positive Psychology: Undisclosed Flexibility in Data Collection and Analysis Allows Presenting Anything as Significant," *Psychological Science* 22, no. 11 (2011): 1359–66, http://journals.sagepub.com

/doi/abs/10.1177/0956797611417632.

On the science journalism side, I recommend: Daniel Engber, "Daryl Bem Proved ESP Is Real: Which Means Science Is Broken," *Slate*, May 17, 2017, https://slate.com/health-and-science/2017/06 /daryl-bem-proved-esp-is-real-showed-science-is -broken.html.

I also owe huge thanks to Kristina Olson (once a mentor, always a mentor!), Simine Vazir (who delivered feedback on a 17-second deadline), and Sanjay Srivastava (whom I cite toward the end of the chapter) for their help and encouragement with this chapter.

- **For a slightly more technical discussion, see the endnotes:** The p-value is the probability of achieving a result at least this extreme, given that the experiment's hypothesis is false.

Or, to unpack it a little more:

1. Assume that we're chasing a fluke, that chocolate doesn't really make people happier.
2. Envision the distribution of all possible results that our experiment could have obtained. Most of these results would be middling, unremarkable, and unlikely to fool us. But a few rare flukes will make it look like chocolate boosts happiness.

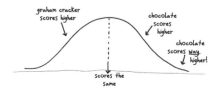

3. Give our actual results a percentile rank in this distribution.

A low score (e.g., 0.03, or the 97th percentile) marks a serious fluke. Only 3% of false results would be this spectacular and deceptive. That extremity suggests that—perhaps—it's not a fluke at all. Maybe the effect we're seeking is real.

Crucially, such evidence is indirect. That 3% isn't the probability of a fluke. It's the probability, *assuming the result is false*, that you'd get a fluke this persuasive.

- **a statistician named R. A. Fisher:** Gerard E. Dallal, "Why P=0.05?" May 22, 2012, http://www .jerrydallal.com/lhsp/p05.htm.
- **Children prefer lucky people:** Kristina Olson et al., "Children's Biased Evaluations of Lucky Versus Unlucky People and Their Social Groups," *Psychological Science* 17, no. 10 (2006): 845–46, http://journals.sagepub.com/doi/abs/10.1111/j.1467-92 80.2006.01792.x#articleCitationDownloadContainer.
- **found the bias across cultures:** Kristina Olson et al., "Judgments of the Lucky Across Development and Culture," *Journal of Personality and Social Psychology* 94, no. 5 (2008): 757–76.
- **the senior thesis of a feckless 21-year-old**: Ben Orlin, "Haves and Have Nots: Do Children Give More to the Lucky Than the Unlucky?" Yale University, senior thesis in psychology, 2009. Adviser: Kristina Olson, who deserves credit for all of the paper's merits and blame for none of its shortcomings.
- **The crucial p-value was well above 0.05**: I predicted that eight-year-olds would be sensitive to

context, giving toys to the unlucky when they had lost a toy, but not when they had experienced an irrelevant form of bad luck (being assigned to play a game with a disliked classmate), whereas the five-year-olds would be insensitive to such contextual issues. The p-value for this test was 0.15.

By comparison, when it came to liking rather than giving, my data replicated Kristina's result, with subjects preferring lucky kids (p = 0.029).

- **the website Spurious Correlations:** Highly recommended: http://www.tylervigen.com/.
- **An anonymous 2011 survey:** Leslie John, George Loewenstein, and Drazen Prelec, "Measuring the Prevalence of Questionable Research Practices with Incentives for Truth Telling," *Psychological Science* 23, no. 5 (2012): 524–32, http://citeseerx.ist .psu.edu/viewdoc/download?doi=10.1.1.727.5139& rep=rep1&type=pdf.
- **Final P-Value vs. Minimum P-Value:** This is a somewhat unfair comparison, because even a p-hacking researcher wouldn't retroactively go back and "stop" the study at the minimum p-value. Indeed, the idea of checking the data after *every* subject is dubious, too.

For a more naturalistic test, I went back to my "types of water" spreadsheet and simulated another 50 trials. This time, I started with 30 subjects (10 per group), then added more subjects as needed in chunks of 15 (five per group), up to a maximum of 90 total. With this extra degree of freedom, a horrifying 76% of trials achieved p-values below 0.05. One trial managed a p-value of 0.00001, which is a 1-in-100,000 absurdity.

- **lowering the p-value threshold:** On first encounter, I thought this sounded stupid. Sort of like: "Our roller coaster is meant for people at least 4 feet tall, but since children are donning trench coats and standing on each other's shoulders to sneak in, let's raise the bar to 5 feet." Then I read the relevant paper: Daniel J. Benjamin et al., "Redefine Statistical Significance," PsyArXiv, July 22, 2017, https:// psyarxiv.com/mky9j/. I'm now persuaded. It seems that dropping from 0.05 to 0.005 would correspond better to intuitive Bayesian thresholds and would reduce false positives while requiring only a moderate increase in sample sizes.

It's also worth noting that Bayesianism is savvier and more sophisticated than I portray it in the chapter text. The question of "What should the prior be?" can be sidestepped by performing numerous analyses, on a variety of priors, and showing a graph of the overall trends. For what it's worth, Sanjay Srivastava persuaded me that a switch to Bayesianism wouldn't really address the causes of the replication crisis, although it might be a good idea for other reasons.

- **careful replications of 100 psychological studies:** Open Science Collaboration, "Estimating the Reproducibility of Psychological Science," *Science* 349, no. 6251 (2015): http://science.sciencemag.org /content/349/6251/aac4716.

Chapter 19: The Scoreboard Wars

- **In 1982, Jay Mathews was the Los Angeles bureau chief:** Jay Mathews, "Jaime Escalante Didn't Just Stand and Deliver. He Changed U.S. Schools Forever," *Washington Post*, April 4, 2010, http://www.washingtonpost.com/wp-dyn/content /article/2010/04/02/AR2010040201518.html.
- **"sheer mockery":** "Mail Call," *Newsweek*, June 15, 2003, http://www.newsweek.com /mail-call-137691.
- **"with the hope that people will argue":** Jay Mathews, "Behind the Rankings: How We Build the List," *Newsweek*, June 7, 2009, http://www.newsweek .com/behind-rankings-how-we-build-list-80725.
- **heed the tale of British ambulances:** Tim Harford, *Messy: How to Be Creative and Resilient in a Tidy-Minded World* (London: Little, Brown, 2016), 171–73.
- **Teacher's Value Added:** This one is real: Cathy O'Neil, *Weapons of Math Destruction: How Big Data Increases Inequality* (New York: Broadway Books, 2016), 135–40.
- **"Nearly every professional educator":** Jay Mathews, "The Challenge Index: Why We Rank America's High Schools," *Washington Post*, May 18, 2008, http://www.washingtonpost.com/wp-dyn /content/article/2008/05/15/AR2008051502741.html.
- **"passer rating":** I've been watching football since I was seven, and have never understood passer rating. I figured it was time to give it a shot.

The formula took a few minutes to unpack, but once I did, I saw that it isn't all that complicated. First, it assigns points (1 per yard, 20 per completion, 80 per touchdown, and -100 per interception). Second, it computes the number of points per passing attempt. And third, it tacks on a meaningless addition/multiplication flourish. Or, in equation form:

$$\text{Passer Rating} = \left(\frac{\text{Yds} + 20 \times \text{Comp.} + 80 \times \text{TDs} - 100 \times \text{Ints}}{\text{Att.}} \right) \times \frac{25}{6} + \frac{25}{12}$$

We could leave it there . . . except this formula allows for negative scores (if you throw too many interceptions) and has an unattainable maximum (whereas the actual maximum of 158⅓ has been achieved in games by more than 60 quarterbacks).

To fix this, you need to cap the points contributed by each stat, and for that, it's easier to present the formula as it is in the main text.

Why does passer rating feel so confusing? As with most scary formulas, two factors come into play: (1) it *is* confusing, being a strangely weighted average of four variables, each subject to arbitrary cutoffs; and (2) on top of that, most sources present it in a needlessly counterintuitive and opaque way. Look on Wikipedia if you want to see what I mean.

- **"One of its strengths is the narrowness":** Mathews, "Behind the Rankings."
- **"The list has taken on a life of its own":** J. P. Gollub et al., eds., "Uses, Misuses, and Unintended Consequences of AP and IB," *Learning and Understanding: Improving Advanced Study of Mathematics and Science in U.S. High Schools*, National Research Council (Washington, DC: National Academy Press, 2002), 187, https://www.nap.edu/read/10129 /chapter/12#187.
- **"The parents are making the most noise":** Steve Farkas and Ann Duffett, "Growing Pains in the Advanced Placement Program: Do Tough Trade-Offs Lie Ahead?" Thomas B. Fordham Institute (2009), http://www.edexcellencemedia.net/publica tions/2009/200904_growingpainsintheadvanced placementprogram/AP_Report.pdf.

According to the survey, more than one in five AP teachers nationwide believed that the Challenge Index had had "some impact" on their school's AP offerings. In suburbs and cities, the number was closer to one in three. Meanwhile, just 17% viewed the list as "a good idea."

- **"Because your index only considers the number":** Valerie Strauss, "Challenging Jay's Challenge Index," *Washington Post*, February 1, 2010, http://voices.washingtonpost.com/answer-sheet /high-school/challenged-by-jays-challenge-i.html.

In 2006, the Challenge Index ranked Eastside High in Gainesville, Florida, as the #6 school in the US. But only 13% of Eastside's nearly 600 African American students were reading at grade level. Schools #21 and #38 showed similar incongruities. Critics saw this as a sign that unprepared students were being shoveled into college-level courses in an effort to catch *Newsweek*'s eye. Source: Michael Winerip, "Odd Math for 'Best High Schools' List,"

New York Times, May 17, 2006, http://www.nytimes
.com/2006/05/17/education/17education.html.

• **Who, exactly, is the index for?:** John Tierney,
"Why High-School Rankings Are Meaningless—and
Harmful," *Atlantic*, May 28, 2013, https://www
.theatlantic.com/national/archive/2013/05/why-high
-school-rankings-are-meaningless-and-harmful
/276122/.

• **"We cannot resist looking at ranked lists":**
Mathews, "Behind the Rankings."

• **"we are all tribal primates":** Jay Mathews,
"I Goofed. But as Usual, a Smart Educator Saved
Me," *Washington Post*, June 25, 2017, https://www
.washingtonpost.com/local/education/i-goofed-but
-as-usual-a-smart-educator-saved-me/2017/06/25
/7c6a05d6-582e-11e7-a204-ad706461fa4f_story
.html.

• **"a very elastic term in our society":** Winerip,
"Odd Math for 'Best High Schools' List."

• **Mathews likes to cite a 2002 study:** Jay
Mathews, "America's Most Challenging High
Schools: A 30-Year Project That Keeps Growing,"
Washington Post, May 3, 2017, https://www.wash
ingtonpost.com/local/education/jays-americas-mo
st-challenging-high-schools-main-column/2017/05/03
/eebf0288-2617-11e7-a1b3-faff0034e2de_story.html.

• **seemed to lay the groundwork for success in
college:** Not all scholars agree. The 2010 book *AP: A
Critical Examination of the Advanced Placement Program* found a consensus building among researchers
that the AP expansion had begun to see diminishing
returns. Coeditor Philip Sadler said, "AP course work
does not magically bestow advantages on under-
prepared students who might be better served by a
course not aimed at garnering college credit." Source:
Rebecca R. Hersher, "The Problematic Growth of
AP Testing," *Harvard Gazette,* September 3, 2010,
https://news.harvard.edu/gazette/story/2010/09
/philip-sadler/.

• **As of 2017, that number is 12%:** Mathews,
"America's Most Challenging High Schools."

• **"the view that schools with lots of rich kids
are good":** Mathews, "The Challenge Index."

• **how you feel about his vision:** I have a humble
suggestion for Mathews: Instead of counting "exams
taken," count "scores of at least 2." In my experience
(and in the Texas study that Mathews likes to cite) a
2 shows some intellectual spark, a signal of growth.
But I'm not convinced the typical 1-scorer has gained
from the class. Requiring students to score above the
minimum would eliminate the incentive to throw
wholly unprepared kids into the test.

Chapter 20: The Book Shredders

• **Ben Blatt's delightful book:** Ben Blatt,
Nabokov's Favorite Word Is Mauve (New York: Simon
& Schuster, 2017).

• **the greatest novels:** I computed the "greatness
score" from data on Goodreads, where readers give
ratings from one to five stars. First, I calculated
the total number of stars awarded to each book.
Faulkner's ranged from under 1500 (for *Pylon*) to
over 500,000 (for *The Sound and the Fury*). Then I
took a logarithm, which collapses this exponential
scale down to a linear one. The correlation between
adverbs and "greatness" was -0.825. Equivalent
analyses for Hemingway and Steinbeck yielded
coefficients of -0.325 and -0.433: substantial, but
hard to discern on a graph. The adverb data comes
from Blatt, and the method is a variation on his. (He
uses number of ratings instead of number of stars; it
yields pretty much the same result.)

• **a blockbuster research article:** Jean-Baptiste
Michel et al., "Quantitative Analysis of Culture
Using Millions of Digitized Books," *Science* 331,
no. 6014 (2011): 176–82, http://science.sciencemag
.org/content/early/2010/12/15/science.1199644.
They wrote:

*The corpus cannot be read by a human. If you
tried to read only the entries from the year 2000
alone, at the reasonable pace of 200 words / minute, without interruptions for food or sleep, it
would take eighty years. The sequence of letters
is one thousand times longer than the human
genome; if you wrote it out in a straight line, it
would reach to the moon and back 10 times over.*

• **"New Window on Culture":** Patricia Cohen, "In
500 Billion Words, New Window on Culture," *New
York Times*, December 16, 2010, http://www.nytimes
.com/2010/12/17/books/17words.html.

• **"precise measurement":** Michel and Aiden
wrote (emphasis mine):

*Reading small collections of carefully chosen
works enables scholars to make powerful
inferences about trends in human thought.
However, this approach rarely enables precise
measurement of the underlying phenomena.*

• **Virginia Woolf's, like a galaxy:** Virginia Woolf,
A Room of One's Own (1929).

• **Apply Magic Sauce:** Try it for yourself at https://
applymagicsauce.com/.

- **25 blog posts:** Taken from http://mathwithbad drawings.com/.
- **In a 2001 paper:** Moshe Koppel, Shlomo Argamon, and Anat Rachel Shimoni, "Automatically Categorizing Written Texts by Author Gender," *Literary and Linguistic Computing* 17, no. 4 (2001): 401-12, http://u.cs.biu.ac.il/~koppel/papers/male -female-llc-final.pdf.
- **A later paper:** Shlomo Argamon et al., "Gender, Genre, and Writing Style in Formal Written Texts," *Text* 23, no. 3 (2003): 321–46, https://www.degruyter .com/view/j/text.1.2003.23.issue-3/text.2003.014 /text.2003.014.xml.
- **As Blatt points out:** Blatt, *Nabokov's Favorite Word Is Mauve*, 37.
- **a data company called CrowdFlower:** Justin Tenuto, "Using Machine Learning to Predict Gender," *CrowdFlower*, November 6, 2015, https://www.crowdflower.com/using-machine-learn ing-to-predict-gender/.
- **gender-indicative words in classic literature:** Blatt, 36.
- **mathematician Cathy O'Neil:** Cathy O'Neil, "Algorithms Can Be Pretty Crude Toward Women," *Bloomberg*, March 24, 2017, https://www.bloomberg.com/view/articles/2017-03-24 /algorithms-can-be-pretty-crude-toward-women.
- **this vindicates Woolf:** In *A Room of One's Own*, Woolf wrote:

> *The weight, the pace, the stride of a man's mind are too unlike [a woman's] for her to lift anything substantial from him . . . Perhaps the first thing she would find, setting pen to paper, was that there was no common sentence ready for her to use.*

Though she enjoys that masculine style ("swift but not slovenly, expressive but not precious"), she added, "It was a sentence that was unsuited for a woman's use."

> *Charlotte Brontë, with all her splendid gift for prose, stumbled and fell with that clumsy weapon in her hands . . . Jane Austen looked at it and laughed at it and devised a perfectly natural, shapely sentence proper for her own use and never departed from it. Thus, with less genius for writing than Charlotte Brontë, she got infinitely more said.*

- **Enter two statisticians:** Frederick Mosteller and David Wallace, "Inference in an Authorship Problem," *Journal of the American Statistical Association* 58, no. 302 (1963): 275–309.
- **tore *The Federalist Papers* to shreds:** I mean that literally. Blatt writes:

> *They took copies of each essay and dissected them, cutting the words apart and arranging them (by hand) in alphabetical order. At one point Mosteller and Wallace wrote, "during this operation a deep breath created a storm of confetti and a permanent enemy."*

I came super close to titling this chapter "A Storm of Confetti and a Permanent Enemy."

- **Stanford Literary Lab:** Sarah Allison et al., "Quantitative Formalism: An Experiment," Stanford Literary Lab, pamphlet 1, January 15, 2011, https:// litlab.stanford.edu/LiteraryLabPamphlet1.pdf. I love this paper. In fact, I recommend every Stanford Literary Lab product I've ever read. They're like vintage Pixar—no bad apples.

Section V: On the Cusp

- **Attempted Light Bulb Designs:** "I have constructed three thousand different theories in connection with the electric light, each one of them reasonable and apparently to be true," Thomas Edison said once. "Yet only in two cases did my experiments prove the truth of my theory." Of course, "theories" aren't the same thing as "design attempts," and no one knows how many designs he tried before success. Source: "Myth Buster: Edison's 10,000 Attempts," *Edisonian* 9 (Fall 2012), http://edison.rutgers.edu /newsletter9.html#4.
- **50 cents, or 51 cents, but not 50.43871 cents:** You can bend this rule if you're doing many transactions at once. Charging $50,438.71 for 100,000 pencils is equivalent to charging $0.5043871 for each. Financial institutions, carrying out tremendous numbers of transactions per day, often work in tiny fractions of pennies.

Chapter 21: The Final Speck of Diamond Dust

- **Adam Smith asked a question:** He didn't really phrase it as a question:

  Nothing is more useful than water: but it will purchase scarce any thing; scarce any thing can be had in exchange for it. A diamond, on the contrary, has scarce any value in use; but a very great quantity of other goods may frequently be had in exchange for it.

  From: Adam Smith, *An Inquiry into the Nature and Causes of the Wealth of Nations* (1776), book I, chapter IV, paragraph 13, accessed through the online Library of Economics and Liberty: http://www.econlib.org/library/Smith/smWN.html.

- **Classical economics lasted a century:** My go-to source for this chapter was this wonderful book: Agnar Sandmo, *Economics Evolving: A History of Economic Thought* (Princeton, NJ: Princeton University Press, 2011).

- **This idea long predates marginalism:** Campbell McConnell, Stanley Brue, and Sean Flynn, *Economics: Principles, Problems, and Policies*, 19th ed. (New York: McGraw-Hill Irwin, 2011). The "earth's fertility" quote comes from section 7.1, Law of Diminishing Returns, http://highered.mheducation.com/sites/0073511447/student_view0/chapter7/origin_of_the_idea.html.

- ***The earth's fertility resembles a spring***: As Mike Thornton points out—and I humbly thank him for his help with this chapter—this analogy elides some subtleties. Heterogeneous land can be perfect for growing various crops that prefer different conditions, and farmers can take steps (such as rotating crops) in order to improve the quality of the soil.

- **This function of utility is peculiar:** William Stanley Jevons, "Brief Account of a General Mathematical Theory of Political Economy," *Journal of the Royal Statistical Society, London* 29 (June 1866): 282–87, https://socialsciences.mcmaster.ca/econ/ugcm/3ll3/jevons/mathem.txt.

- **"A true theory of economy":** Jevons again. Quotable guy. Writing this chapter made me a fan.

- **In school, I learned that economics has mirror symmetry:** For a refresher, I drew from the lecture notes of Michal Brzezinski, an assistant professor of economics at the University of Warsaw, http://coin.wne.uw.edu.pl/mbrzezinski/teaching/HEeng/Slides/marginalism.pdf.

- **self-described "stand-up economist":** Yoram Bauman. I quote from his video *Mankiw's Ten Principles of Economics, Translated*, which cracked me up when I first saw it in college. Check out his site at http://standupeconomist.com/.

- **"The proper mathematical analogy":** Sandmo, *Economics Evolving*, 96.

- **"persist in using everyday language":** Sandmo, 194.

- **"the only work by an economist":** The quote is from Wikipedia. Shhh, don't tell anyone.

- **renewed empiricism:** Jevons embodied this trend. He predicted that Britain would soon exhaust its coal resources, and argued that the ups and downs of the business cycle stemmed from cooler temperatures caused by sunspots. Okay, so he was wrong about both, but we know that thanks to methods that he himself ushered in.

  Walras, by contrast, was a bit of an anti-empiricist. In his view, you must first filter out the messy details of reality to arrive at pure quantitative concepts. Second, you must operate on and reason about these mathematical abstractions. Third, almost as an afterthought, you look back to practical applications. "The return to reality," Walras wrote, "should not take place until the science is completed." To Walras, "science" was something that happens far away from reality.

  If you meet an economist who still thinks this way today, just follow these simple steps: (1) raise your arms; (2) begin bellowing; (3) if the economist continues to approach, box him on the ears. Remember, these economists are as frightened of us as we are of them.

- **Economics, like its brethren in the natural sciences, became ahistorical:** This insight comes straight from Sandmo.

Chapter 22: Bracketology

- **"Famous historical Belgians":** I'm not kidding. Belgium's two greatest celebrities are both fictional: Hercule Poirot and Tintin. Personally, I count this as a point in Belgium's favor. Famous historical people are nothing but trouble.
- **Our tale begins in 1861:** My source for the history of US taxation was: W. Elliot Brownlee, *Federal Taxation in America: A History*, 3rd ed. (Cambridge: Cambridge University Press, 2016).
- **you now owed 3%:** The cutoff point of $800 landed quite close to the nation's average family income of $900 (per Brownlee). Pretty impressive, given that the government threw that particular dart blindfolded. I should also note that the taxes didn't really have a 0% bracket; rather, they had a universal exemption of $800. The math is equivalent.
- **In 1913, the first year of the permanent income tax:** Brownlee, *Federal Taxation in America*, 92.
- **President Woodrow Wilson wrote:** Brownlee, 90.
- **the 1918 tax scheme:** Numbers drawn from the Wikipedia page "Revenue Act of 1918."
- **One industrious fellow named J. J.:** J. J.'s system requires calculus to pull off, which goes to show that J. J. was not your typical precalculus student. His system (unlike the simpler one I illustrate) had several discrete rate jumps leading into a continu-ously varying top bracket, whose formula involved a natural logarithm. To be a teacher often involves simplifying the world so your students can grasp it; here, it seems to entail the reverse.
- **Donald Duck:** *The New Spirit* (Walt Disney Studios, 1942). To see the magnificent original, go to https://www.youtube.com/watch?v=eMU -KGKK6q8.
- **their greatest album:** Yes, *Revolver* is their best. Come at me, *Rubber Soul* fans.
- ***one for you, nineteen for me:*** This corresponds to a marginal rate of 95%, so George Harrison was slightly *under*stating his tax burden.
- **a woman named Pomperipossa:** Astrid Lindgren, "Pomperipossa in Monismania," *Expressen*, March 3, 1976. I used Lennart Bilén's helpful 2009 translation: https://lenbilen.com/2012/01/24 /pomperipossa-in-monismania/.
- **caught fire in Sweden:** "Influencing Public Opinion," AstridLindgren.se, accessed September 2017, http://www.astridlindgren.se/en/person /influencing-public-opinion.
- **higher income levels faced a *lower* marginal tax rate:** Of course, this is how the actual payroll tax, designed to fund Social Security and Medicare, works. You pay 15% on your first $120,000, and nothing on income above that level.

Chapter 23: One State, Two State, Red State, Blue State

- **democracy is an act of mathematics:** After writing this cheerful line, I came across its evil twin. My beloved Jorge Luis Borges, in a moment of uncharacteristic cynicism, once called democracy "an abuse of statistics."
- **"men chosen by the people":** *Federalist*, no. 68 (Alexander Hamilton). While I'm name-dropping beloved American figures, I shall thank Geoff Koslig (a man so constitutionally patriotic that his brain's background music is "You're a Grand Old Flag") for his help with this chapter. (I'd also like to thank Jeremy Kun and Zhi Chen for their capable help.)
- **Who Chose Electors?:** I drew this line of argument, like 93% of my knowledge about the Constitution, from Akhil Reed Amar. In particular, from this book: Akhil Reed Amar, *America's Constitution: A Biography* (New York: Random House, 2005), 148–52. The data in the graph comes from (where else?) Wikipedia.
- **this system encoded an advantage for slavery:** This argument comes from Amar, *America's Constitution*, 155–59, 342–47. He goes further, arguing that the Electoral College not only advantaged slave states (which is pretty indisputable) but was designed in part to do so. This position has drawn some criticism, including Gary L. Gregg II's energetically titled: "No! The Electoral College Was Not about Slavery!" *Law and Liberty*, January 3, 2017, http://www.libertylawsite.org/2017/01/03 /no-the-electoral-college-was-not-about-slavery/.

My $0.02: It would be silly to argue that slavery was the *only* factor that shaped the Electoral College. But I don't see anyone saying that. On page 155, Amar writes: "three main factors—information barriers, federalism, and slavery—doomed direct presidential election in 1787." Meanwhile, the idea that slavery was a *non*factor strikes my amateur eyes as quite wrong. On July 19, 1787, James Madison raised the issue:

> There was one difficulty . . . attending an
> immediate choice by the people. The right
> of suffrage was much more diffusive in the
> Northern than the Southern States; and the
> latter could have no influence in the election

on the score of the Negroes. The substitution of electors obviated this difficulty . . .

On July 25, Madison raised the point again—although this time, he endorsed direct election as his favorite method, saying: "As an individual from the S. States he was willing to make the sacrifice." Read the details for yourself in "Notes on the Debates in the Federal Convention": http://avalon .law.yale.edu/subject_menus/debcont.asp.

Further evidence that the slavery advantage was a feature, not a bug: in his first inaugural address in 1833, Southerner Andrew Jackson proposed replacing the Electoral College with a direct election—but only if the irregularities in vote weighting were preserved. Source: Amar, 347.

• **Here's the elector count from 1800:** I drew census data from Wikipedia. Then, I redistributed the number of representatives in the House in proportion to the free population of each state.

• **bucket-filled with red and blue:** Koslig informed me that the "Democrat = blue, Republican = red" color scheme was once the opposite. It shifted in the 1990s and was cemented in 2000. For more, you can read a fun *Smithsonian* piece: Jodi Enda, "When Republicans Were Blue and Democrats Were Red," Smithsonian.com, October 31, 2012, https://www.smithsonianmag.com/history /when-republicans-were-blue-and-democrats -were-red-104176297/.

• **60% of the vote gets six electors:** Even this is not straightforward as it sounds. Take the 2016 election, where the Democrats' 46.44% rounds up to 50% (for five electors), and the Republicans' 44.92% rounds down to 40% (for four electors). That's only nine. Who gets the last elector? You could give it

to the Libertarians, but their 3.84% is pretty far from 10%. It's probably more logical to give the last elector to the party that was closest to earning it outright—in this case, the Republicans, who were just 0.08% away.

• **where 96% of the country sits today:** As Koslig points out, this dynamic is different in states that favor one party in presidential elections and the other for state government. In recent years, a few such states have considered joining Nebraska and Maine in their way of allocating electors, as a way for the state-government-ruling party to seize some electors from its rival.

• **a beautiful method for answering this question:** Nate Silver, "Will the Electoral College Doom the Democrats Again?" FiveThirtyEight, November 14, 2016, https://fivethirtyeight.com /features/will-the-electoral-college-doom-the -democrats-again/.

• **no correlation between which party has the Electoral College advantage:** "The major exception," Silver points out,

was in the first half of the 20th century, when Republicans persistently had an Electoral College advantage because Democrats racked up huge margins in the South, yielding a lot of wasted votes. . . . The question is whether Democrats are re-entering something akin to the "Solid South" era, except with their votes concentrated in more urban coastal states . . . .

Silver's guess seems to be "no." Time will tell.

• **the Amar Plan:** It originated with Professor Akhil Reed Amar (who taught the superb Constitutional Law class I took in college) and his brother, Professor Vikram Amar.

Chapter 24: The Chaos of History

• **8 inches of snow fell on the eve of President Kennedy's inauguration**: Source: Wikipedia, of course. I also take thank David Klumpp and Valur Gunnarsson for brain-tickling feedback on this chapter.

• **a researcher named Edward Lorenz found something funny**: This anecdote, the ensuing graph, and indeed most of the mathematics in this chapter come from the indispensable: James Gleick, *Chaos: Making a New Science* (New York: Viking Press, 1987). The anecdote is found on pages 17–18.

• **The pendulum obeyed a simple equation:** Okay, they didn't have "meters" in the 17th century. This approximation is a modern restatement, and

it holds only for fairly small-angle swings. Still, it's striking that the cycle length doesn't depend on the angle: a wide, fast oscillation will take about the same time as a slower, narrower one.

• **The double pendulum:** For an awesome simulator, check out: https://www.myphysicslab.com /pendulum/double-pendulum-en.html. Seriously, if you've been skimming the endnotes in search of some unknown inspiration, then this is it.

• **The Game of Life:** Siobhan Roberts, *Genius at Play: The Curious Mind of John Horton Conway* (New York: Bloomsbury, 2015). On pages xiv–xv, Roberts quotes musician Brian Eno on the Game of Life: "The whole system is so transparent that there should be no surprises at all, but in fact there are plenty: the complexity and 'organicness' of the evolution of the dot patterns completely beggars prediction." On page 160, she quotes philosopher Daniel Dennett: "I think that Life should be a thinking tool in everybody's kit."

• **"The big choices we make":** This Amos Tversky quote comes from: Michael Lewis, *The Undoing Project: A Friendship That Changed Our Minds* (New York: W. W. Norton, 2016), 101.

• **Kim Stanley Robinson's novella:** Kim Stanley Robinson, *The Lucky Strike* (1984). Excellent story. I read it in: Harry Turtledove, ed., *The Best Alternate History Stories of the 20th Century* (New York: Random House, 2002).

Also, I can't help mentioning here my favorite alternate-history novel: Orson Scott Card, *Pastwatch: The Redemption of Christopher Columbus* (New York: Tor Books, 1996).

• **Kyoto had remained atop the target list:** Mariko Oi, "The Man Who Saved Kyoto from the Atomic Bomb," *BBC News*, August 9, 2015, http://www.bbc.com/news/world-asia-33755182.

• **Hiroshima was a civilian target:** Hard to imagine that Truman didn't understand this, but it seems he didn't. Listen to the excellent episode: "Nukes," *Radiolab*, April 7, 2017, http://www.radiolab.org/story/nukes.

• **Winston Churchill, not primarily known for his science fiction:** In his 1931 essay "If Lee Had NOT Won the Battle of Gettysburg," Churchill pretends that he is a historian living in an alternate timeline, where the Confederacy won; then he writes an "alternate" fiction imagining what life is like in *our* timeline. I found his conclusions stupid and naïve, but then again, I never saved a civilization from Nazis, so don't feel compelled to listen to me. https://www.winstonchurchill.org/publications/finest-hour-extras/qif-lee-had-not-won-the-battle-of-gettysburgq/

• **"*Confederate* is a shockingly unoriginal idea":** Ta-Nehisi Coates, "The Lost Cause Rides Again," *Atlantic*, August 4, 2017, https://www.theatlantic.com/entertainment/archive/2017/08/no-confederate/535512/.

• **The rules of *plausibility*:** This discussion was inspired in part by Michael Lewis's *The Undoing Project*, 299–305.

• **shed light on the current state:** A key insight of chaos theory is that many systems of towering complexity follow simple underlying rules. Maybe— and here I slip into the realm of late-night dorm-room speculation—there's a way to articulate a few deterministic rules that underlie (aspects of) history. We might develop toy models to capture history's sensitivity and degree of randomness, the way Edward Lorenz's simulation did for the weather.

• **Benoit Mandelbrot published a brief bombshell of a paper:** Benoit Mandelbrot, "How Long Is the Coast of Britain? Statistical Self-Similarity and Fractional Dimension," *Science* 156, no. 3775 (1967): 636–38.

• **"Each new book":** Alan Moore and Eddie Campbell, *From Hell* (Marietta, GA: Top Shelf Productions, 1999), appendix II, 23.

• **something bottomless about the study of history:** In Ursula Le Guin's science fiction story "A Man of the People," the protagonist seeks to understand the sprawling history of a civilization called Hain:

> *He knew now that historians did not study history. No human mind could encompass the history of Hain: three million years of it . . . . there had been uncountable kings, empires, inventions, billions of lives lived in millions of countries, monarchies, democracies, oligarchies, anarchies, ages of chaos and ages of order, pantheon upon pantheon of gods, infinite wars and times of peace, incessant discoveries and forgettings, innumerable horrors and triumphs, an endless repetition of unceasing novelty. What is the use trying to describe the flowing of a river at any one moment, and then at the next moment, and then at the next, and the next, and the next? You wear out. You say: There is a great river, and it flows through this land, and we have named it History.*

Gosh, do I love this passage. Source: Ursula Le Guin, *Four Ways to Forgiveness* (New York: HarperCollins, 1995), 124–25.

• **three PowerPoint presentations projected onto a single screen:** I stole this nifty image from Valur Gunnarsson.

Also, on the topic of competing historical metaphors, I can't help quoting Borges: "It may be that universal history is the history of the different intonations given a handful of metaphors."

ACKNOWLEDGMENTS

The elements of this book are a bit like the atoms in my body: only nominally and temporarily "mine." They've circulated for years, coming from too many sources to credit or trace. The best I can do is gesture toward the whole ecosystem that made this book possible.

For the book's style, I thank all of the witty, kindhearted folks at the *Yale Record*, with special hat tips to Davids Klumpp and Litt, and Michaels Gerber and Thornton.

For the book's perspective, I thank my extraordinary colleagues at King Edward's School, with special affection for Tom, Ed, James, Caz, Richard, Nei—ah, heck, for everybody. Teachers are a playful, critical, inclusive, curious, and slightly batty tribe that I'm proud to call my people.

For the book's sense of purpose, I thank my students and my teachers, who have shaped in uncountable ($\aleph_1$) ways my thinking about math and about the world.

For the book's errors (especially any omitted thank-yous), I apologize in advance.

For the book's existence, I thank: the dozens of gracious people who gave feedback and advice (see the endnotes); Chank Diesel, for elegantly AutoTuning my handwriting into something radio-ready; Mike Olivo, for educating me about wookiees; Paul Kepple, for assembling so many bad drawings into a beautiful whole; Elizabeth Johnson, for bringing peace to the ugly war between myself and hyphens, and for knowing how many *a*'s are in "*Daaaaaamn*"; Betsy Hulsebosch, Kara Thornton, and the rest of the Black Dog & Leventhal team; Dado Derviskadic and Steve Troha, for being the first to envision this book—eons before I could—and for helping me to get there; and Becky Koh, for her wonderful editing, a job that seems to combine the trickiest elements of Executive Producing and parenting.

Love and gratitude to my family: Jim, Jenna, Caroline, Lark, Farid, Justin, Diane, Carl, my happy triangle Soraya, my potato wizard Skander, Peggy, Paul, Kaya, and the whole Orlin, Hogan, and Williams clans. With fond memories of Alden, Ros, Pauline, and, of course, Donna.

Finally: thanks, Taryn. You picked math and I'm glad I came along for the ride. I love you even more than I troll you.